Contents

The Chicago Handbook of University Technology Transfer and Academic Entrepreneurship

The Chicago Handbook of University Technology Transfer and Academic Entrepreneurship

Edited by
ALBERT N. LINK, DONALD S. SIEGEL, AND MIKE WRIGHT

THE UNIVERSITY OF CHICAGO PRESS CHICAGO AND LONDON

ALBERT N. LINK is professor of economics at the University of North Carolina at Greensboro. DONALD S. SIEGEL is dean of the School of Business and professor of management at the University at Albany, the State University of New York. MIKE WRIGHT is professor of entrepreneurship and head of the Department of Innovation and Entrepreneurship at Imperial College Business School where he also serves as associate director of the Enterprise Research Center and director of the Centre for Management Buy-out and Private Equity Research.

The University of Chicago Press, Chicago 60637
The University of Chicago Press, Ltd., London

Printed in the United States of America

24 23 22 21 20 19 18 17 16 15   1 2 3 4 5

ISBN-13: 978-0-226-17834-9 (cloth)
ISBN-13: 978-0-226-17848-6 (e-book)
DOI: 10.7208/chicago/9780226178486.001.0001

Library of Congress Cataloging-in-Publication Data

The Chicago handbook of university technology transfer and academic entrepreneurship / [edited by] Albert N. Link, Donald S. Siegel, and Mike Wright.
pages ; cm
Includes index.
ISBN 978-0-226-17834-9 (cloth : alk. paper)—ISBN 978-0-226-17848-6 (e-book)
1. Academic spin-outs. 2. Technology transfer. 3. University-based new business enterprises. 4. Technology transfer—United States. 5. Academic spin-outs—United States. 6. University-based new business enterprises—United States. I. Link, Albert N., editor. II. Siegel, Donald S., 1959– editor. III. Wright, Mike, 1952– editor.
T174.3.C4787 2015
338.9′26—dc23

2014016929

♾ This paper meets the requirements of ANSI/NISO Z39.48-1992 (Permanence of Paper).

Editors' Introduction

Albert N. Link, Donald S. Siegel, and Mike Wright

Introduction

In recent years, there has been a substantial increase in the rate of technology commercialization at universities. In 1980, Congress passed the Bayh-Dole Act, the landmark legislation governing university technology transfer. Bayh-Dole dramatically changed incentives for universities and firms to engage in university technology transfer. It simplified the commercialization process by instituting a uniform patent policy and removing many restrictions on licensing federally funded technologies. More importantly, it enabled universities to own the patents that arise from federal research grants. Almost all universities now have a technology transfer or licensing office, and several have also established venture units (e.g., Boston University and Columbia University), as well as proof of concept centers (e.g., Iowa State University and Syracuse University). Also, universities are increasingly being viewed by policy makers as engines of economic growth and local/regional development, via their commercialization of intellectual property through technology transfer.

Formal technology transfer mechanisms include patents, copyrights, trademarks, licensing agreements between the university and private firms, and university-based start-ups, as well as property-based institutions such as incubators and accelerators and research, science, and technology parks. There is also a burgeoning interdisciplinary scholarly literature on university technology transfer, as indicated by the many conferences and special issues of academic journals on this topic. A concomitant trend is the growing number of graduate courses and programs related to technological entrepreneurship in business, engineering, law, and public policy schools, as well as in the social sciences (e.g., economics, political science/public administration, and sociology).

In the aftermath of the Bayh-Dole Act, additional supporting legislation (e.g., the development of the Small Business Innovation Research Program in the United States and ProTon Europe, a European network of technology transfer offices and companies interacting with universities), and an increase in public-private, university-industry research partnerships (e.g., SEMATECH), led to a rapid increase in technology commercialization at universities. Universities are now in the business of managing intellectual property portfolios and are often aggressively attempting to commercialize discoveries from their laboratories. This activity is partially driven by a potentially lucrative stream of licensing revenue (e.g., Columbia University earned over $178 million in licensing revenue in 2003) and IPO-related wealth resulting from Internet search engines and browsers (e.g., Stanford University earned $336 million from the sale of its Google stock in 2005), Gatorade, gene sequencing, and drug discovery. Universities have also been compelled, especially in the United Kingdom and continental Europe, to aggressively pursue technology commercialization, owing to shrinking endowments, reductions in government funding, and increasing operating costs.

Unfortunately, formal management of an intellectual property portfolio is still a relatively new phenomenon for many universities. This responsibility has led to considerable uncertainty among administrators and policy makers governing institutions of higher education regarding optimal organizational practices relating to inventor incentives, technology transfer "pricing," legal issues, strategic objectives, and measurement and monitoring mechanisms.

The rise in university technology transfer has also attracted considerable attention in the academic literature. While many authors have analyzed the antecedents and consequences of university patenting and licensing, some researchers have also assessed the entrepreneurial dimensions of university technology transfer. These authors have analyzed new *institutions* that have emerged to facilitate commercialization, such as university technology transfer offices (henceforth, TTOs); industry-university cooperative research centers (IUCRCs); engineering research centers (ERCs); incubators and accelerators; and research, science, and technology parks. Other research focused more directly on *agents* involved in technology commercialization, such as academic scientists and entrepreneurs. Specifically, numerous scholars have examined the determinants and outcomes of faculty involvement in university technology transfer, such as their propensity to patent, disclose inventions, coauthor with industry scientists, and form university-based start-ups.

There has also been a substantial increase in public investment in research, science, and technology parks; incubators; accelerators; and other property-based institutions that facilitate the transfer of technology from universities to firms. While licensing has traditionally been the most popular mechanism for commercialization of university-based intellectual property, universities are increasingly emphasizing the creation of new companies. National and state governments have introduced a variety of measures to support university technology transfer, including subsidized funding schemes.

Such trends have diffused internationally into varying university regimes. Differences across countries include contexts where ownership of intellectual property varies between the academic researcher and the university's administration, as well as differences in the ability and incentives for academics to engage in technology transfer. These developments have critical economic, organizational, and societal implications, which this handbook seeks to address. The purpose of this handbook is to integrate and synthesize academic studies of university technology transfer and academic entrepreneurship. In doing so, it focuses on both the managerial and policy implications of the rise of this activity in advanced industrial nations.

It is important to note that this handbook is the first definitive source of major academic research on university technology transfer. Given that the literature on university technology transfer is highly interdisciplinary, another important aspect of this handbook is that the contributors represent a variety of social sciences (e.g., economics, sociology, psychology, and political science), fields in business administration (e.g., strategy, organizational behavior, entrepreneurship, marketing, and finance), and other professional programs and areas of study (e.g., law, public administration, and engineering). Since university technology transfer is a global phenomenon, the handbook includes a substantial amount of international evidence, which reflects a variety of national perspectives on this topic.

The Contributions

In the remainder of this introduction, we summarize each chapter and attempt to draw some lessons learned for university administrators; technology transfer office directors and personnel; managers at firms; incubators; accelerators; research, science, and technology parks; and policy makers. The key research question and findings of each contribution are summarized in table 0.1.

Siegel and Wright review and synthesize recent studies of the antecedents and consequences of university technology transfer, presenting some lessons learned for university administrators, policy makers, and corporations and entrepreneurs interacting with universities. They note that several scholars have found that a key concern, from the perspective of university administrators, is that many academics are failing to disclose inventions to the TTO. The authors also comment on the importance of royalty regimes, inventor involvement, and equity stakes in commercializing technology. The dual agency problem in TTOs emphasizes the need for appropriate governance and incentives inside TTOs.

The evidence reviewed shows that differences among universities in commercialization activities are closely connected to environmental and institutional factors within universities, indicating the importance of developing organizational practices at the university to promote commercialization. Although it is difficult to generalize across different types of institutions, key "best practices" include reducing cultural barriers among academics, university administrators, and firms/entrepreneurs, encouraging faculty involvement in technology transfer through more favorable royalty distribution formulas, incorporation of technology transfer activities into promotion and tenure criteria, and implementing incentive-based compensation within TTOs. Universities also need to manage tensions among traditional activities (e.g., research, teaching, and service) and those related to technology transfer, by establishing ambidextrous organizational structures and processes.

There is also a need to upgrade personnel skills and capabilities in TTOs and to reconfigure offices into smaller units, possibly with a regionally based sectorial focus. The extent of new venture creation based on university intellectual property (IP) is highly influenced by human capital (e.g., staffing of TTOs, "star scientists," and entrepreneurial teams) and the university culture (e.g., university research strength, department

TABLE 0.1. **Chapters in *The Chicago Handbook of University Technology Transfer and Academic Entrepreneurship.***

Authors	Research Question	Key Results
Donald Siegel (University at Albany, SUNY) and Mike Wright (Imperial College London)	What Are the Most Effective Organizational Practices Relating to University Technology Transfer Offices, Licensing, and Start-Ups?	Key Organizational Practices to Stimulate More Effective University Technology Transfer: 1) Faculty-Friendly Royalty Distribution Formulas 2) Reward Technology Transfer in Promotion/ Tenure 3) Incentive-Based Compensation for TTO Staff 4) Establish Ambidextrous Organizational Structures and Processes 5) Upgrade TTO Skills and Capabilities 6) Bring Entrepreneurs to Campus 7) Formulate a Technology Transfer Strategy
Markus Perkmann (Imperial College London) and Joel West (Keck Graduate Institute)	Examines the Relationship between Open Innovation and University Technology Transfer (from the Perspective of the Firm)	Finds that the Three Modes of Interaction between Firms and Universities- IP Licensing, Research Services, and Research Partnerships-Are Compatible with Open Innovation
Arti Rai (Duke University) and Bhaven Sampat (Columbia University)	Analysis of the "Public Interest" Provision of the Bayh-Dole Act	Reviews the Somewhat Sparse Academic Literature on Compliance with Reporting Provisions of Bayh-Dole Act. The Authors Also Conduct Their Own Quantitative Analysis of the Biomedical Arena. They Find Evidence of Noncompliance, Especially in the Area of Drug Discovery
Fiona Murray (MIT) and Julian Kolev (MIT)	Examines University Technology Transfer from the Perspective of the Entrepreneur	Successful Commercialization Requires that Both Internal and External Entrepreneurs Understand the Three Distinct "Levels" of University Technology Transfer: (1) The *National* Level and *Institutional* Rules and the Legal Environment for Commercialization; (2) the *Local* Level Involving Local Interpretation and Implementation of the Bayh-Dole Act and (3) the *Individual* Level Involving the Social Networks, Norms, and Incentives for Different Actors Involved in Commercialization
Andrew Nelson (University of Oregon) and Tom Byers (Stanford University)	A Comprehensive Analysis of the Entrepreneurial Dimension of University Technology Transfer, Including the Role of Entrepreneurship Education	Considers the Resource Requirements for Successful Start-Ups, Including Financial Resources, Facilities, Specialized Equipment, and Human Capital (i.e., Managers, Team Members, Board Members, and Advisors). Also Considers the Role of Entrepreneurship Education in Addressing These Resource Requirements. More Importantly, the Authors Demonstrate How Entrepreneurship Education Can Enhance Technology Transfer, but Caution against Excessively Close Relationships and the Cooptation of Entrepreneurship Education for Technology Transfer Aims

(*continued*)

TABLE 0.1. *(continued)*

Authors	Research Question	Key Results
Albert Link (University of North Carolina at Greensboro) and John Scott (Dartmouth College)	How Do Research, Science, and Technology Parks Influence Knowledge Flows and Technology Transfer?	Reviews the Literature on Research, Science, and Technology Parks; Concludes that Parks Play a Significant Role in Stimulating Knowledge Flows and Technology Transfer between Universities and Firms and Thus, Deserve Greater Support
David Audretsch (Indiana University) and Devrim Göktepe-Hulten (Lund University)	Why Is University Technology Transfer Less Active in Europe than in the U.S.?	Comprehensive Review of University Patenting in Europe; Faculty Ownership of Intellectual Property ("Professor's Privilege") Doesn't Impede European Technology Transfer Performance, Which Is Better than Previously Reported; This Mis-measurement Is Due to the Fact that in Europe, There Is More *Informal* University Technology Transfer than in the U.S.
Nicola Baldini (University of Bologna), Riccardo Fini (Imperial College London) and Rosa Grimaldi (University of Bologna)	Comprehensive Study of University Technology Transfer/Academic Entrepreneurship in Italy	The Authors Assess Academic Entrepreneurship at 64 Science, Technology, Engineering and Mathematics Universities in Italy during the Last Decade; Universities with the Following Characteristics Have Better Results: 1) Greater Autonomy in the Management of Intellectual Property Rights 2) Those with TTO Personnel Who Have Been Trained for Technology Transfer 3) Universities with Support Mechanisms for Academic Entrepreneurship
Tucker Marion (Northeastern University), Denise Dunlap (Northeastern University), and John Friar (Northeastern University)	What Are the Key Determinants of Successful Academic Entrepreneurship?	The Most Important Determinants of a Successful University Spin-out Are: (1) A Faculty Member's Level of Entrepreneurial Experience; (2) Inclination toward Commercialization; (3) Participation in an Industry Sponsored Research Agreement. Universities Interested in Bolstering Research-Based New Ventures Should Actively Recruit and Retain Faculty with More Entrepreneurial Experience, Reward Academic Entrepreneurship in Promotion and Tenure, and Promote Entrepreneurial Learning and Thinking through Training

chairs, and entrepreneurs who are employed at these institutions) and group norms. The growth of spin-offs that are created presents major challenges that include the need to develop external commercial networks and to evolve boards and management teams with commercial and prior entrepreneurial experience. Additional policy conclusions are that universities and regions must formulate a coherent and feasible technology transfer strategy that takes account of their research strengths, the balance between different modes of commercialization and the concomitant IP strategies, and collaborations with other universities. The authors also discuss several controversial issues related to TTOs, such as whether some universities constitute "patent trolls" and whether commercialization and entrepreneurship should be rewarded in promotion and tenure decisions. Finally, the role of business schools in university technology transfer and academic entrepreneurship is also considered.

Perkmann and West examine the role of open science and open innovation in the sourcing of knowledge by firms, through their direct and formal interactions with universities. They observe that although in some sectors universities are a key source of knowledge and innovation for firms, research on open innovation has paid relatively scant attention to university-industry relations. Therefore, they review the broader literature on firm-university interactions from the firm's viewpoint, identify how it informs our knowledge of this specific variant of open innovation, and then suggest opportunities for further research that integrates these two literatures.

Perkmann and West provide an overview of three main modes of direct interaction between firms and universities: IP licensing, research services, and research partnerships. They outline the main characteristics of each mode, their relative importance for firms, and benefits and challenges. While licensing remains an important mode by which public-funded, university-based research is used by firms, they highlight the significant role of relationship-based modes of interaction—research services and research partnerships—between firms and universities. The authors conclude that some of these relationship-based interactions enable appropriation via intellectual property rights, while others are more aligned with the norms of open science and create benefits for firms by generating basic knowledge, creating skills, and enabling follow-on innovation. They suggest that despite different emphases, there is no necessary contradiction between any of these modes of interaction and open science. Technology transfer via licensing occurs when the science has already been conducted, and in many cases research results have been published in the open literature in parallel to being patented. The creation of intellectual property rights often arises from work conducted during research partnerships and research services, and research services are often provided by academics whose inventions have been licensed to firms but require more inventor involvement to successfully exploit the technologies concerned. Finally, they suggest that there is a need for further research to understand the process of open source knowledge acquisition from the perspective of the firm to balance what we already know from the vantage point of the university.

Rai and Sampat examine the public interest provisions of the Bayh-Dole Act. A key aspect of protecting the public interest, in terms of this legislation, was a set of compliance and reporting provisions. These include the requirements to disclose government interests in federally funded patents and to report these patents to funding agencies. The authors begin by discussing the U.S. Department of Commerce's implementation regulations and available case law on sanctions for noncompliance with reporting requirements. They conclude that there is little empirical evidence on compliance. The authors then review large-scale quantitative data in the biomedical area to obtain insight into the implementation of reporting requirements and suggest there is evidence of noncompliance. Finally, they undertake a qualitative review focused on drug development, an area where accountability is arguably especially important, and provide specific evidence regarding why they believe federal funding was not appropriately reported for at least 15 patents (associated with eight drugs). They also show that for five other patents (associated with five drugs), federal funding was involved in at least a portion of the relevant research or in research on compounds very closely related to the patented drug. Overall, this evidence is suggestive of compromising the underlying value of patents.

Murray and Kolev examine university technology transfer from the perspective of entrepreneurs who are external and internal to the university. Focusing on the U.S. context, they examine how entrepreneurs can engage effectively with universities in this process. The authors conclude that they can do so by understanding the legal rules shaping engagement in commercialization, local rules and norms, and the incentives and expectations that guide faculty attitudes toward entrepreneurs. Based on both systematic research of the university and its role in commercialization, as well as on anecdotal evidence from MIT and other universities, they argue that both internal and external entrepreneurs need to be aware of and understand three distinct levels of technology transfer if they are to be successful in the commercialization process. These are (1) the national level involving the Bayh-Dole Act and the legal environment for commercialization; (2) the local level involving each institution's interpretation and implementation of the national-level rules; and (3) the individual level involving the social networks and norms together with the incentives for the different actors involved in commercialization. They view the structure of the relationship between entrepreneurs and researchers as central to effective commercialization.

Nelson and Byers outline the many challenges that confront efforts to commercialize university technologies via spin-outs or start-ups. They also consider the role of entrepreneurship education, in relation to these challenges. The authors begin by considering the role of start-ups vis-à-vis other mechanisms in the commercialization of university research. Next, they outline the resource requirements for successful start-ups, including financial resources; facilities; specialized equipment; and people, including potential managers, team members, board members, and advisors. This is followed by a consideration of the role of entrepreneurship education in addressing these resource requirements. Drawing upon an extensive literature review, they elaborate on best practices for entrepreneurship education, in terms of audience, curriculum, and external engagement. In their concluding section, they highlight a number of important distinctions between entrepreneurship education and technology transfer, and put forth a set of questions that can aid programs in assessing the relationship between these areas. Ultimately, they point to a number of ways by which entrepreneurship education can enhance technology transfer, but caution against excessively close relationships and the cooptation of entrepreneurship education for technology transfer aims.

Link and Scott review the literature on knowledge and technology transfer at research, science, and technology (R-S-T) parks and consider the public policy implications of such findings. The authors conclude that successful two-way knowledge flows between universities and industry are a key ingredient of a "national innovation system" and that R-S-T parks play a significant role in stimulating such flows. However, they note that parks are not a sine qua non of these knowledge flows. The authors suggest that while R-S-T parks fall under the broader category of an effective educational system, they warrant a higher status in the future, as technological life cycles continue to shorten and as basic research at universities and applied research/development in industry become more intertwined.

Audretsch and Göktepe-Hultén revisit the notion of the so-called European academic paradox. This starts from the observation that the level of commercial activity at European universities is relatively low, compared to the high levels of scientific performance and investment in research. It is alleged that an important determinant of this "paradox" lies in the existence of the "professor's privilege," which refers to the fact that in some European countries, academics, rather than their universities, own the patents based on the faculty member's research. The authors build on studies that suggest that the paradox might not necessarily exist, given that differences in the number of university patents can be explained by the differences in legislations and that abolition of the professor's privilege may be unnecessary and futile. In particular, they address to what extent the law of professor's privilege has fostered university inventions to be pre-emptied by incumbent firms and decreased the likelihood and intentions of entrepreneurship among academics. They show that the lack of university patents in Europe is actually a lack of university-owned patents, not necessarily a lack of university-invented patents.

Once the data are corrected to take into account the different ownership structure in Europe and the U.S., their calculations suggest that the European academic system seems to perform much better than previously believed. The authors then discuss whether this law causes a tendency toward "solution-oriented research" at universities, and only contributes to in-house research activities of existing firms, rather than science-driven research. They argue that the law of professor's privilege facilitates relations with industrial partners, since scientists can easily transfer the ownership rights of their intellectual property to the partner firms, without the intervention of TTOs. The professor's privilege and close informal relations with existing firms, as well as the contractual structure of competence centers, lead to a tendency for a reverse linear model of innovation, where research questions are generated by industrial needs. They show that the nature of university patenting in Europe depends not only on the ownership regimes but also on the informal and formal collaborative patterns of university-industrial relations.

Their comprehensive review of key empirical studies reveals that most university technology transfer activity has been informal, and undertaken by large firms without universities claiming any rights over these patents. Finally, the authors relate their discussion to whether the critical view on TTOs is grounded on the argument that incumbent companies' monopoly on the direct appropriation of university knowledge will be challenged. They report evidence that academic inventors find it difficult to accept the role and involvement of TTOs, since this is viewed as impractical and unnecessary as the firms they collaborate with are the entities that apply for the patents. Academic investors are also critical of the contribution of TTOs, yet it may be that established firms may gain access to the most attractive technology. The authors conclude that while politicians and university administrators focus on legislation and organizational structures for academic entrepreneurship, the main component of entrepreneurship—that is, research—has not received the attention and investment it needs. While science parks and incubators are important nurturers of university knowledge, they need a constant infusing of new knowledge generated through research.

Baldini, Fini, and Grimaldi present a comprehensive study of university technology transfer and academic entrepreneurship in Italy, which is a relatively recent phenomenon in that country. Specifically, they assess the extent of academic entrepreneurship at 64 science, technology, engineering, and mathematical universities during the previous decade. Their findings indicate that universities possessing the following characteristics tend to have better results in university technology transfer:

1) Institutions having greater autonomy in the management of intellectual property rights;
2) Those employing TTO personnel who have been trained specifically for technology transfer;
3) Universities with support mechanisms for academic entrepreneurship.

Marion, Dunlap, and Friar analyze academic entrepreneurship, in order to identify factors that correlate with success in launching a new venture. The authors conduct an empirical analysis of 400 disclosures over a ten-year period at Northeastern University in Boston, Massachusetts. They find that the most important determinants of successful university spin-outs are a faculty member's level of entrepreneurial experience, inclination toward commercialization, and participation in an industry-sponsored research agreement. Their results imply that if universities are interested in bolstering research-based new ventures, they should actively recruit faculty with more entrepreneurial experience and be entrepreneurial themselves. To attract and retain such academic entrepreneurs, universities may need to broaden the scope of activities from recruiting requirements to rewards and promotion, including incorporating entrepreneurial achievements into promotion and tenure policies, and promoting entrepreneurial learning and thinking through training.

CHAPTER ONE

University Technology Transfer Offices, Licensing, and Start-Ups

Donald S. Siegel and Mike Wright

1. Introduction

In recent decades, almost all research universities in the U.S. and Europe have established technology transfer offices (henceforth, TTOs) to commercialize their intellectual property. In the U.S., the Association of University Technology Managers (AUTM 2013) reports that the annual number of patents granted to U.S. universities rose from less than 300 in 1980 to 5,145 in 2012, while licensing of new technologies has increased almost sixfold since 1991. Annual licensing revenue generated by U.S. universities rose from about $160 million in 1991 to $2.6 billion in 2012. University-based start-up companies numbering 705 were launched in 2005 alone; while 6,834 new firms based on university-owned intellectual property have been created since 1980.

The pattern observed in the U.S. is part of an international phenomenon, with substantial increases in licensing, patenting, and university-based start-up companies also evident across Europe, Australia, Canada, and elsewhere (Wright et al. 2007). Technologies that have been commercialized via a TTO include the famous Boyer-Cohen "gene-splicing" technique that launched the biotechnology industry, diagnostic tests for breast cancer and osteoporosis, Internet search engines (e.g., Lycos), music synthesizers, computer-aided design (CAD), and environmentally friendly technologies.

TTOs serve as an "intermediary" between suppliers of innovations (university scientists) and those who can potentially (help to) commercialize these innovations (i.e., firms, entrepreneurs, and venture capitalists). TTOs facilitate commercial knowledge transfers of intellectual property resulting from university research through licensing to existing firms or start-up companies of inventions or other forms. The activities of TTOs have important economic and policy implications. On the positive side, licensing agreements and university-based start-ups (spin-offs) can result in additional revenue for the university, employment opportunities for university-based researchers and graduate students, and local economic and technological spillovers through the stimulation of additional R&D investment and job creation. On the negative side, the costs of TTOs may outweigh the revenues generated and there may be other deleterious effects, such as the potential diversion away from research activity with longer-term pay-offs.

There is increasing recognition that technology commercialization can occur through several modes. Traditionally, the emphasis has been on licensing and patenting. Thursby and Thursby (2007) examine the policy issues relating to university licensing. More recently, increased attention has been devoted to the creation of spin-off firms by academic scientists. Scholars have examined university technology commercialization and entrepreneurship, typically focusing on the "performance" of TTOs, while also analyzing agents engaged in commercialization, such as academic scientists. Several authors have assessed the determinants and outcomes of faculty involvement in technology commercialization, such as the propensity of academics to patent, disclose inventions, coauthor with industry scientists, and form university-based start-ups.

These developments give rise to a number of important policy-related questions:

1. What is the role of TTOs in commercializing university intellectual property?
2. How successful are TTOs in generating revenue from the various modes of university intellectual property?
3. What are the key challenges in enabling TTOs to generate revenues from university intellectual property
4. What policies need to be developed to meet these challenges?

In this chapter, we review this burgeoning literature and derive some lessons learned for practitioners and policy makers. The remainder of this chapter is organized as follows. In section 2, we describe the role of TTOs and review some key theoretical studies. Next, we discuss empirical studies on the efficiency of TTOs in commercializing university IP. Both the evidence on university licensing and patenting (section 3) and university-based start-ups (section 4) are examined. Section 5 presents lessons learned and policy recommendations.

2. Theoretical Analysis on the Role of the TTO

Consistent with Siegel, Waldman, and Link (2003), we conjecture that there are three agents involved in commercialization: university scientists, technology transfer or licensing officers and/or other university research administrators, and corporate (venture capital) managers and/or entrepreneurs who (help to) commercialize university-based technologies.

Differences in Culture and Objectives

In understanding the commercialization process, it is useful to reflect on the incentives and cultures of these three key agents.

Firms and entrepreneurs seek to commercialize university-based intellectual property for profit. When innovation is a key source of competitive advantage, it is critical to maintain proprietary control over these technologies. Therefore, firms typically wish to secure exclusive rights to university-based technologies. Speed is another major area of concern, since firms and entrepreneurs often seek to establish a "first-mover" advantage.

The *TTO and other university administrators* generally regard themselves as the guardians of the university's intellectual property portfolio, which can potentially generate revenue. Therefore, they are anxious to market university-based technologies to companies and entrepreneurs, although they will often "hedge," since they do not want to be accused of "giving away" lucrative taxpayer-funded technologies or because they want to safeguard the "researchers and the research environment" that generates innovations. This tends to slow down the commercialization process.

Academic scientists, especially those who are untenured, seek the rapid dissemination of their ideas and breakthroughs. This propagation of new knowledge is manifested along several dimensions, including publications in the most selective scholarly journals, presentations at leading conferences, and research grants. The end result of such activity is peer recognition, through citations and stronger connections to the key social networks in academia. Faculty members may also seek pecuniary rewards, which can be pocketed or ploughed back into their research to pay for laboratory equipment, graduate students, and postdocs. Lacetera (2009) discusses the decision by academic research teams to undertake commercially oriented activities and their subsequent performance, as compared to industrial teams, taking into account the differences in objectives and organizational structure they face. Aghion, Dewatripont, and Stein (2008) model the specific characteristics of agents belonging to the scientific community, as compared to industrial teams when discussing the commercialization decision. These models are helpful in explaining differential performance of academic versus nonacademic spin-offs and the decision when to license versus spinning off.

Problems in Commercialization of University Intellectual Property

University scientists are the suppliers of new innovations, in the sense that they discover new knowledge while conducting (funded) research projects. However, before a university-based innovation can be commercialized, several hurdles must be surmounted. Key issues are whether researchers have sufficient incentives to disclose their inventions, how to induce researchers' cooperation in further development in bringing IP to market, and whether it is possible to overcome asymmetric information problems relating to the value of university inventions.

DISCLOSURE OF INVENTIONS AND OWNERSHIP OF IP. Invention disclosures to the university constitute the critical "input" in the technology transfer process. In the U.S., the Bayh-Dole Act requires that academics who are being funded on a federal research grant disclose their inventions to the university/TTO. Across Europe, there are notable differences in the ownership of university IP, although it is possible to discern some convergence. In the U.K., universities have increasingly enforced their ownership rights to IP generated by academic scientists, with the royalties associated with it being distributed between the relevant parties on an institutionally organized basis. Germany and Belgium adopted Bayh-Dole type legislation in the late 1990s, while in France this type of regulation had existed for some time. In Italy, public researchers receive the right to ownership of their IP, but in most cases the university makes a formal contract on an individual basis to give the IP rights to the university. Discussions are in progress in Sweden and Finland to change to a Bayh-Dole type arrangement from a model of inventor ownership. Most European countries have changed their legislation to make it possible and/or more attractive to researchers and academics to assume equity in a start-up and/or receive royalties. In France, for example, it was illegal before 1999 for an academic to assume equity in a start-up company (for additional discussion of this issue, see Wright et al. 2007).

It is important to note that the process of university technology transfer should begin, in theory, with an invention disclosure. However, based on extensive interviews with academic scientists in the U.S., Siegel et al. (2004) reported that many faculty members are not disclosing their inventions to their universities. Survey research by Thursby, Jensen, and Thursby (2001) confirms this finding. Markman, Gianiodis, and Phan (2008) have documented that many technologies are indeed "going out the back door."

The failure of many academics to disclose inventions to the TTO highlights the problems for officers in eliciting disclosures. Although the Bayh-Dole Act stipulates that scientists must file an invention disclosure, this rule is rarely enforced. Instead, the university needs to have proper incentive schemes in place, specifying an adequate share for the inventors in royalties or equity. The importance of this share in securing researchers' cooperation in technology licensing has been analyzed by Macho-Stadler, Martinez-Giralt, and Pérez-Castrillo (1996); Lach and Schankerman (2004); Link and Siegel (2005); and Jensen and Thursby (2001). All of these models focus on licensing, rather than commercialization through start-ups. Nevertheless, empirical studies of start-up formation by universities have demonstrated the importance of royalty regimes of the university, even on academic spin-off creation rates (e.g., Di Gregorio and Shane 2003; and O'Shea, Allen, and Chevalier 2005).

BRINGING UNIVERSITY IP TO THE MARKET. If a faculty member files an invention disclosure, the TTO (which, in most countries, is the owner of the university IP) decides whether the invention should be patented, usually in consultation with a committee of faculty experts. In making this decision, the TTO typically attempts to assess the commercial potential of the invention. Sometimes, firms or entrepreneurs have already expressed sufficient interest in the new technology to warrant filing a patent. If industry expresses little interest in the technology, universities may be reluctant to file for a patent, given the high cost of filing and protecting patents. When a patent is filed and awarded, the university typically attempts to "market" the invention, by contacting firms that can potentially license the technology or entrepreneurs who are capable of launching a start-up firm based on the technology.

Faculty members may also become directly involved in the licensing agreement as technical consultants or as entrepreneurs in a university spin-off. Jensen and Thursby (2001) show that faculty involvement in the commercialization of a licensed university-based technology increases the likelihood that such an effort will be successful. In order to provide incentives for faculty involvement, licensing agreements should entail either upfront royalties or royalties at a later date. For spin-offs, Macho-Stadler, Pérez-Castrillo, and Veugelers (2008) demonstrate how the optimal contract between the university TTO, the researcher, and the venture capitalist entails the allocation to the researcher of an equity stake to secure her involvement in the venture. It may also require the researcher to be financially involved in the project as a way to give her incentives to provide effort. The creation and development of a spin-off may generate higher financial returns to universities than licensing. However, as discussed in the next section of this chapter, TTOs may encounter numerous challenges in pursuing entrepreneurial activities.

ASYMMETRIC INFORMATION AND THE VALUATION OF INVENTIONS. Even when the disclosure and researcher involvement problem is mitigated by an appropriate incentive scheme, not all potentially viable inventions will be patented and licensed by the university. This relates to the problem of asymmetric information between industry and academia regarding the value of an invention. Firms typically cannot assess the quality of the invention ex ante, while researchers may find it difficult to assess the commercial profitability of their inventions. This problem is examined in Macho-Stadler, Pérez-Castrillo, and Veugelers (2007), who use a reputation argument for a TTO to alleviate the asymmetric information problem. The authors demonstrate that larger TTOs may have an incentive to shelve some projects, in order to build a reputation for delivering good projects, thus raising buyers' beliefs of expected quality. Their results support the importance of a critical size for TTOs to be successful. The authors also predict that establishing a TTO may result in fewer license agreements, but higher average license revenues. In contrast, Clarysse et al. (2007) demonstrate that the problem may be that TTOs are perceived as placing too high a value on the patent or innovation, in part because they are incentivized to maximize income generation, while buyers and investors are unwilling to meet this valuation because of the uncertainty of generating future income streams from it.

The Rationale for TTOs

Several theoretical papers have provided rationales for universities to establish TTOs. For Macho-Stadler, Pérez-Castrillo, and Veugelers (2007), the rationale for establishing a TTO is in the pooling of inventions for reputation building. In contrast, Jensen, Thursby, and Thursby (2003) model the process of faculty disclosure and university licensing in the U.S. through a TTO as a game, in which the principal is the university administration and the faculty and TTO are agents who maximize expected utility. This approach recognizes the greater complexity in the principal-agent relationships in universities as compared to firms and is in line with recent developments in multiple agency theory (Arthurs et al. 2008; Bruton et al. 2010). The authors treat the TTO as a dual agent—that is, as an agent of both the faculty and the university.

While there are vertical principal-agent relationships between the university and the TTO and between the university and the faculty, the relationship between the TTO and the faculty is more of a horizontal agency relationship. Faculty members must decide whether to disclose the invention to the TTO and at what stage—that is, whether to disclose at the most embryonic stage or wait until it is a lab-scale prototype. The university administration influences the incentives of the TTO and faculty members by establishing university-wide policies for the shares of licensing income and/or sponsored research. If an invention is disclosed, the TTO decides whether to search for a firm to license the technology and then negotiates the terms of the licensing agreement with the licensee. Quality is incorporated in their model as a determinant of the probability of successful commercialization. According to the authors, the TTO engages in a "balancing act," in the sense that it can influence the rate of invention disclosures, must evaluate the inventions once they are disclosed, and negotiates licensing agreements with firms as the agent of the administration. However, as Clarysse et al. (2007) show, the dual agent role of TTOs and their need to convince their own boards of their achievements in attracting investment funds may lead them, as noted above, to overvalue projects compared to their underlying quality. All this calls for proper governance and incentive schemes inside the TTO (see also Belenzon and Schankerman [2009]).

Hellman (2005) models the advantage a TTO has, compared to individual scientists or teams, in terms of lower search costs for potential buyers, owing to specialization and/or lower opportunity cost of time. He finds that scientists are more likely to delegate their search to TTOs when there is patent protection. Similarly, Hoppe and Ozdenoren (2005) present a theoretical model to explore the conditions under which innovation intermediaries, like TTOs, emerge to reduce the uncertainty problem. In their model, firms seek to invest in inventions, but they cannot estimate the value of the technology with certainty. Intermediaries are able to incur a sunk investment, acquiring the expertise to locate new inventions, sort profitable ones from unprofitable ones, and assess the efficiency level of potential licensees. The authors show that the fixed start-up costs of TTOs can be recovered if the size of the invention pool is large enough to exploit the economies of sharing expertise. The intermediary will reduce the uncertainty problem, although the authors still find a high probability of inefficient outcomes resulting from coordination failure.

As noted in Siegel and Phan (2005), TTOs are increasingly encountering a key strategic choice regarding the commercialization of intellectual property: whether to emphasize licensing or spin-offs. Such choices are likely to be determined by the university's perception of the relative financial returns and their desire to generate economic/knowledge spillovers to the community. Significant informational asymmetries and uncertainties regarding the potential markets for innovations may affect this choice. TTOs may have an advantage over academic scientists in identifying opportunities and developing spin-offs through their commercial networks and business development expertise. However, scientists can play a hold-up role if their expertise is necessary for the development of the technology and their preference is for a spin-off rather than a licensing agreement (Lockett et al. 2005). Furthermore, attracting and creating appropriate incentives for TTO personnel, who possess the appropriate set of skills required to develop spin-offs, rather than negotiating licenses, can be problematic.

The stark disparities in the motives, perspectives, and cultures of the three key players in this process underscore the potential importance of understanding how organizational factors and institutional policies influence effective university management of intellectual property. In the following two sections of the paper, we review empirical studies of the institutional and managerial practices involved in the commercialization of university intellectual property. The next section considers studies relating to licensing and patents, while section 4 addresses start-ups.

3. Effectiveness of TTOs: Licensing and Patents

In table 1.1, we present a review of selected empirical studies of the performance of TTOs with respect to licensing and patenting. Some of these papers attempt to evaluate the productivity of TTOs, based on data measuring the "outputs" and "inputs" of university technology transfer (e.g., Siegel, Waldman, and Link 2003; and Thursby and Thursby 2002). These studies are typically based on a production function framework, in which a "best practice" frontier is constructed. The distance from the frontier represents the level of "technical" inefficiency, or the inability of the organization to generate maximal output from a given set of inputs. Two methods are used to estimate these frontiers-data envelopment analysis (DEA) and stochastic frontier estimation (SFE).

Thursby and Kemp (2002) and Thursby and Thursby (2002) employ data envelopment analysis (DEA) methods to assess whether the growth in licensing and patenting by universities can be attributed to an increase in the willingness of professors to patent, without a concomitant, fundamental change in the type of research they conduct. The alternative hypothesis is that the growth in technology commercialization at universities reflects a shift away from basic research toward more applied research. The authors find support for the former hypothesis. More specifically, they conclude that the rise in university technology transfer is the result of a greater willingness on the part of university researchers to patent their inventions, as well as an increase in outsourcing of R&D by firms via licensing.

Siegel, Waldman, and Link (2003) use stochastic frontier estimation (SFE) to pose a different research question: Why are some universities more effective at transferring technologies than comparable institutions? Specifically, they attempt to assess and "explain" the relative productivity of 113 U.S. university TTOs. Contrary to conventional economic models, they found that variation in relative TTO performance

TABLE I.I. **Selected Papers on University Technology Transfer Offices.**

Author(s)	Data Sets	Methodology	Key Results
Siegel, Waldman, and Link (2003)	AUTM, NSF, and U.S. Census Data, Interviews	TFP of University Licensing -Stochastic Frontier Analysis and Field Interviews	TTOs Exhibit Constant Returns to Scale with Respect to the # of Licensing; Increasing Returns to Scale with Respect to Licensing Revenue; Organizational and Environmental Factors Have Considerable Explanatory Power
Link and Siegel (2005)	AUTM, NSF, and U.S. Census Data, Interviews	TFP of University Licensing -Stochastic Frontier Analysis	Land Grant Universities Are More Efficient in Technology Transfer; Higher Royalty Shares for Faculty Members Are Associated with Greater Licensing Income
Friedman and Silberman (2002)	AUTM, NSF, NRC, Milken Institute "Tech-Pole" Data	Regression Analysis-Systems Equations Estimation	Higher Royalty Shares for Faculty Members Are Associated with Greater Licensing Income
Lach and Schankerman (2004)	AUTM, NSF, NRC	Regression Analysis	Higher Royalty Shares for Faculty Members Are Associated with Greater Licensing Income
Rogers, Yin and Hoffmann (2000)	AUTM, NSF, NRC	Correlation Analysis of Composite Technology Transfer Score	Positive Correlation Between Faculty Quality, Age of TTO, and # of TTO Staff and Higher Levels of Performance in Technology Transfer
Thursby, Jensen, and Thursby (2001)	AUTM, Authors' Survey	Descriptive Analysis of Authors' Survey/ Regression Analysis	Inventions Tend to Disclosed at an Early Stage of Development; Elasticities of Licenses and Royalties with Respect to Invention Disclosures Are Both Less than One; Faculty Members Are Increasingly Likely to Disclose Inventions.
Foltz, Barham and Kim (2000)	AUTM, NSF	Linear Regression	Faculty Quality, Federal Research Funding, and # of TTO Staff Have a Positive Impact on University Patenting
Bercovitz, Feldman, Feller, and Burton (2001)	AUTM and Case Studies, Interviews	Qualitative and Quantitative Analysis	Analysis of Different Organization Structures for Technology Transfer at Duke, Johns Hopkins, and Penn State; Differences in Structure May Be Related to Technology Transfer Performance
Thursby and Kemp (2002)	AUTM	Data Envelopment Analysis and Logit Regressions on Efficiency Scores	Faculty Quality and # of TTO Staff Has a Positive Impact on Various Technology Transfer Outputs; Private Universities Appear to Be More Efficient than Public Universities; Universities with Medical Schools Less Efficient

Author(s)	Data Sets	Methodology	Key Results
Thursby and Thursby (2002)	AUTM and Authors' Own Survey	Data Envelopment Analysis	Growth in University Licensing and Patenting Can Be Attributed to an Increase in the Willingness of Professors to Patent and License, As Well As Outsourcing of R&D by Firms; Not to a Shift toward More Applied Research
Chapple, Lockett, Siegel, and Wright (2005)	U.K.-NUBS/ UNICO Survey-ONS	Data Envelopment Analysis and Stochastic Frontier Analysis	U.K. TTOs Exhibit Decreasing Returns to Scale and Low Levels of Absolute Efficiency; Organizational and Environmental Factors Have Considerable Explanatory Power
Carlsson and Fridh (2002)	AUTM	Linear Regression	Research Expenditure, Invention Disclosures, and Age of TTO Have a Positive Impact on University Patenting and Licensing

cannot be completely explained by environmental and institutional factors. The implication of this finding is that organizational practices are likely to be an important determinant of relative performance.

Based on estimates of their "marginal product," it appears that technology licensing officers add significant value to the commercialization process. The findings also imply that spending more on lawyers reduces the number of licensing agreements, but increases licensing revenue. Licensing revenue is subject to increasing returns, while licensing agreements are characterized by constant returns to scale.

The authors supplemented their econometric analysis with qualitative evidence, derived from 55 structured, in-person interviews of 100 university technology transfer stakeholders at five research universities in Arizona and North Carolina. The field research allowed them to identify intellectual property policies and organizational practices that can potentially enhance technology transfer performance.

The qualitative analysis identified three key impediments to effective university technology transfer. The first was informational and cultural barriers between universities and firms, especially for small firms. Another impediment was insufficient rewards for faculty involvement in university technology transfer. This includes both pecuniary and nonpecuniary rewards, such as credit toward tenure and promotion. Some respondents even suggested that involvement in technology transfer might be detrimental to their career. Finally, there appear to be problems with staffing and compensation practices in the TTO. One such problem is a high rate of turnover among licensing officers, which is detrimental toward the establishment of long-term relationships with firms and entrepreneurs. Other concerns are insufficient business and marketing experience in the TTO, and the possible need for incentive compensation.

In a subsequent paper, Link and Siegel (2005) report that a particular organizational practice can potentially enhance technology licensing: the "royalty distribution formula," which stipulates the fraction of revenue from a licensing transaction that is allocated to a faculty member who develops the new technology. The authors find that universities allocating a higher percentage of royalty payments to faculty members tend to be more efficient in technology transfer activities. Organizational incentives for university technology transfer appear to be important. This finding was independently confirmed in Friedman and Silberman (2003) and Lach and Schankerman (2004), using slightly different methods and data. Jensen, Thursby, and Thursby (2003) provide an extensive survey of the objectives, characteristics, and outcomes of licensing activity at 62 U.S. universities.[1] They also find that faculty quality is positively associated with the rate of invention disclosure at the earliest stage of idea development and negatively associated with the share of licensing income allocated to inventors. Having incentive schemes in place works not only through eliciting higher efforts of researchers, but also by attracting more productive researchers.

Chandler (1962) and Williamson (1975) drew attention to the economics of internal organization. Bercovitz et al. (2001) build on this framework to examine what could be a critical implementation issue in university management of technology transfer: the *organizational structure* of the TTO and its relationship to the overall university research administration. They analyze the performance implications of four organizational forms: the functional or unitary form (U-Form), the multidivisional (M-form), the holding company (H-form), and the matrix form (MX-form). The authors note that these structures have different implications for the ability of a university to coordinate activity, facilitate internal and external information flows, and align incentives in a manner that is consistent with its strategic goals with respect to technology transfer. To test these assertions, they examine TTOs at Duke, Johns Hopkins, and Penn State and find evidence of alternative organizational forms at these three institutions. They attempt to link these differences in structure to variation in technology transfer performance along three dimensions: transaction output, the ability to coordinate licensing and sponsored research activities, and incentive alignment capability.

Ambos et al. (2008) compare university projects that generate commercial outcomes (patents, licenses, or spin-outs) with those that do not, at the point of the decision regarding whether to commercialize or not. Consistent with ambidexterity theory, they find that universities manage the tensions between academic and commercial demands by setting dual structures. Further evidence emphasizes the importance of the process and content of the interactions between different parts of universities in contributing the knowledge required to exploit innovations. Specifically, universities' management structures may hinder the ability of TTOs, science departments, and business schools to interact with each other to promote technology commercialization (Wright et al. 2009). Differences in cultures, language, and codes among these units of universities may also necessitate the development of boundary spanning roles to facilitate the interaction. The culture of academic departments and the head of department's attitudes toward commercialization may also influence whether faculty in that department engage in commercialization activity whatever the stance of the university itself (Bercovitz and Feldman 2008; Rasmussen, Mosey, and Wright 2014). While additional research is needed to make conclusive statements regarding organizational structure and performance, these findings imply that organizational form of the TTO and its relationship with other parts of the university do matter.

Belenzon and Schankerman (2009) study the use of incentive schemes within the TTO to tie TTO managers' incentives to overall university objectives. They find that bonuses raise licensing income by increasing the quality of transacted inventions, while specifying local development objectives or imposing government constraints on licensing practices of TTOs generally have a negative impact on licensing revenues.

Other papers have focused exclusively on the corporate perspective on formal university technology transfer. Hertzfeld, Link, and Vonortas (2006) interviewed and then surveyed chief intellectual property attorneys at 54 R&D-intensive U.S. firms concerning intellectual property protection mechanisms related to university patents. They found that firms expressed great difficulty in dealing with university TTOs on intellectual property issues, citing the inexperience of the TTO staff, the TTO's lack of general business knowledge and its tendency to overstate the commercial value of the patent. The authors reported that firms had decided to by-pass the TTO in some cases and deal directly with the university scientist or engineer.

Most empirical studies of the performance of TTOs have been based on U.S. data. In recent years, several papers based on E.U. data have been published. Using DEA and SFE methods, Chapple et al. (2005) find in a study of 50 U.K. universities that TTOs exhibit a low level of absolute efficiency in licensing activity and that there appeared to be decreasing returns to scale. These findings indicate that the recent growth in the size of TTOs has not been accompanied by a corresponding growth in the business skills and capabilities of TTO managers. The findings imply a need to upgrade skills and capabilities and to reconfigure TTOs into smaller units, possibly with a regionally based sector focus.

Debackere and Veugelers (2005) explore the case of K.U. Leuven R&D, the technology transfer organization affiliated with K.U. Leuven in Belgium, as well as a comparison group of 11 European research universities. They identify numerous factors influencing the management of technology transfer relationships. Consistent with evidence from the U.S. (see Link and Siegel 2005), they find that incentives and organization practices are important, in terms of explaining variation in relative performance. Specifically, they report that universities allocating a higher percentage of royalty payments to faculty members tend to be more effective in technology transfer. On the organizational side, the authors find that another critical success factor is what they call a "decentralized management style," which apparently allows the technology transfer office to be much more sensitive to the needs of its stakeholders.

In sum, the extant literature on TTOs suggests that the key impediments to effective university technology transfer tend to be organizational in nature (Siegel, Waldman, and Link 2003; Siegel et al. 2003). These include problems with differences in organizational cultures between universities and (small) firms, incentive structures including both pecuniary and non-pecuniary rewards, such as credit toward tenure and promotion, and staffing and compensation practices of the TTO itself.

4. Effectiveness of TTOs: Start-Ups

TTOs are increasingly focused on the start-up dimension of university technology transfer.[2] This increase in start-up or spin-off activity has begun to attract considerable attention in the academic literature. Some studies of entrepreneurial activity use the TTO or university as the unit of analysis, while others focus on individual entrepreneurs. Each aspect is important in understanding the challenges involved in realizing the revenue generating potential of start-ups.

University Characteristics and Start-Up Formation

University culture and strategy may influence the extent of start-up activity and resources devoted to their development. Based on a qualitative analysis of five European universities that had outstanding performance in technology transfer, Clarke (1998) concluded that the existence of an entrepreneurial culture at those institutions was a critical factor in their success. Roberts (1991) finds that social norms and MIT's tacit approval of entrepreneurs were critical determinants of successful academic entrepreneurship at MIT.

Universities that generate the most start-ups have clear, well-defined strategies regarding the formation and management of spin-offs (Lockett, Wright, and Franklin 2003). Degroof and Roberts (2004) and Roberts and Malone (1996) find that the optimal university policy is a comprehensive selectivity/support policy for generating start-ups that can exploit ventures with high growth potential. However, such a comprehensive policy is an ideal that may not be feasible, given resource constraints. The authors conclude that while spin-off policies do matter, in the sense that they affect the growth potential of ventures, it may be more desirable to formulate such policies at a higher level of aggregation than the university. Roberts and Malone (1996) conjecture that Stanford generated fewer start-ups than comparable institutions in the early 1990s because the institution refused to sign exclusive licenses to inventor founders.

As with licensing, the organization and use of resources is an important influence on the start-up creation and development process. Based on evidence from 50 European universities, Clarysse et al. (2005) identify three optimal cases where universities can adopt different resource and capability commitments depending upon whether they are seeking to create small numbers of start-ups that become global businesses generating significant capital gains, national-level businesses generating revenue streams, or a larger number of smaller consultancy and service businesses generating local employment. They also identify suboptimal

TABLE I.2. **Selected Papers on Start-Up Formation at Universities.**

Author(s)	Unit of Analysis	Data/Methodology	Key Results
Di Gregorio and Shane (2003)	University-Based Start-Ups	AUTM Survey/Count Regressions of the Determinants of the # of Start-Ups	Two Key Determinants of Start-Up Formation: Faculty Quality and the Ability of the University and Inventor(s) to Take Equity in a Start-Up, in Lieu of Licensing Royalty Fees; A Royalty Distribution Formula that Is More Favorable to Faculty Members Reduces Start-Up Formation
O'Shea, Allen, and Arnaud (2004)	University-Based Start-Ups	AUTM Survey/Count Regressions of the Determinants of the # of Start-Ups	A University's Previous Success in Technology Transfer Is a Key Determinant of Its Rate of Start-Up Formation
Franklin, Wright, and Lockett (2001)	TTOs and University-Based Start-Ups	Authors' Quantitative Survey of U.K. TTOs	Universities that Wish to Launch Successful Technology Transfer Start-Ups Should Employ a Combination of Academic and Surrogate Entrepreneurship
Lockett, Wright, and Franklin, (2003)	TTOs and University-Based Start-Ups	Quantitative and Qualitative Surveys of U.K. TTOs	Universities that Generate the Most Start-Ups Have Clear, Well-Defined Spin-Off Strategies, Strong Expertise in Entrepreneurship, and Vast Social Networks
Lockett and Wright (2005)	TTOs and University-Based Start-Ups	Survey of U.K. TTOs/ Count Regressions of the Determinants of the # of Start-Ups	A University's Rate of Start-Up Formation is Positively Associated with Its Expenditure on Intellectual Property Protection, the Business Development Capabilities of TTOs, and the Extent to which Its Royalty Distribution Formula Favors Faculty Members
Clarysse, Wright, Lockett, van de Elde and Vohora (2005)	TTOS and University Based Start-Ups	Interviews and Descriptive Data on 50 Universities across 7 European Countries	Five Incubation Models Identified. Three Match Resources, Activities, and Objectives: Low Selective, Supportive and Incubator. Two Do Not: Competence Deficient and Resource Deficient
Markman, Phan, Balkin, and Gianiodis (2004b)	TTOs and University-Based Start-Ups	AUTM Survey, Authors' Survey/ Linear Regression Analysis	There Are Three Key Determinants of Time-to Market (Speed): TTO Resources, Competency in Identifying Licensees, and Participation of Faculty-Inventors in the Licensing Process
Markman, Phan, Balkin, and Gianiodis (2005)	TTOs and University Start-Ups	AUTM Survey, Authors' Survey/ Linear Regression Analysis	The Most Attractive Combinations of Technology Stage and Licensing Strategy for New Venture Creation—Early Stage Technology and Licensing for Equity—Are *Least* Likely to Favored by the University (Due to Risk Aversion and a Focus on Short-Run Revenue Maximization)

Author(s)	Unit of Analysis	Data/Methodology	Key Results
Bercovitz and Feldman (2004)	Medical School Researchers at Johns Hopkins and Duke	Determinants of the Probability of Filing an Invention Disclosure	Three Factors Influence the Decision to Disclose Inventions: Norms at the Institutions Where the Researchers Were Trained and the Disclosure Behaviors of Their Department Chairs and Peers
Audretsch (2000)	Entrepreneurs in the Life Sciences	101 Founders of 52 Biotech Firms/ Hazard Function Regression Analysis	Academic Entrepreneurs Tend to Be Older, More Scientifically Experienced
Louis, Blumenthal, Gluck, and Stoto (1989)	Faculty Members in the Life Sciences	778 Faculty Members from 40 Universities/ Regression Analysis	Key Determinant of Faculty-Based Entrepreneurship: Local Group Norms; University Policies and Structures Have Little Effect
Lowe and Gonzalez (2007)	Faculty Members	150 Faculty Members from 15 Universities/ Regression Analysis	Faculty Members Are More Productive Researchers than Observationally Equivalent Colleagues *before* They Established Their Firms. The Research Productivity of These Academics Did *Not* Decline in the Aftermath of Their Entrepreneurial Activity
Zucker, Darby, and Brewer (1998)	Relationships Involving "Star" Scientists and U.S. Biotech Firms	Scientific Papers, Data on Biotech Firms from the North Carolina Biotechnology Center (1992) and Bioscan (1993)/ Count Regressions	Location of Star Scientists Predicts Firm Entry in Biotechnology
Zucker, Darby, and Armstrong (2000)	Relationships Involving "Star" Scientists and U.S. Biotech Firms	Scientific Papers Reporting Genetic-Sequence Discoveries, Count Regressions	Collaboration between Star Scientists and Firm Scientists Enhances Research Performance of U.S. Biotech Firms, as Measured Using Three Proxies: # of Patents Granted, # of Products in Development, and # of Products on the Market
Zucker and Darby (2001)	Relationships Involving "Star" Scientists and Japanese Biotech Firms	Data on Biotechnology Firms and the Nikkei Biotechnology Directory	Collaboration between Star Scientists and Firm Scientists Enhances Research Performance of Japanese Biotech Firms, as Measured Using Three Proxies: # of Patents Granted, # of Products in Development, and # of Products on the Market

(*continued*)

TABLE 1.2. *(continued)*

Author(s)	Unit of Analysis	Data/Methodology	Key Results
Vanaelst, Clarysse, Wright, Lockett, S'Jegers (2006)	Start-Ups and Entrepreneurial Team Members	Interview Data and Comparative Univariate Analysis	The Spin-Off Team's Heteroegeneity Changes as the Firms Evolve; the Privileged Witness Plays a Key Advisory Role in the Early Stages; New Team Members Enter with Different Experience but not Different View on Doing Business
Kenney and Patton (2011)	University Spin-Offs	Hand-Collected Census of Spin-Offs from Six Universities	Inventor Ownership Universities Are More Efficient in Generating Spin-Offs; More Spin-Offs Are Generated in Computer Science and Electrical Engineering than in Biomedical or Physical Sciences.
Wennberg, Wiklund, and Wright (2011)	University Spin-Offs and Corporate Spin-Offs	Population of 8,663 CSOs and 528 U.S.Os in Swedish Knowledge Intensive Sectors 1994–2002	Corporate Spin-Offs, Especially Those Involving University Graduates Who Gained Industrial Experience, Perform Better than University Spin-Offs

cases where university TTOs either attempt to undertake commercialization activities beyond the resources they have available or where they fail to develop the competences needed to achieve particular commercialization targets. These findings indicate a need to match universities' objectives for commercialization, and for spin-offs in particular, with appropriate resources and capabilities and with a realistic perspective on the types of successful spin-offs that can be created given a particular university's science and resource base.

The specific expertise and track record of TTOs may also influence the creation and economic development of spin-offs from universities. O'Shea, Allen, and Chevalier (2005) find that a university's previous success in technology transfer is a key explanatory factor of start-up formation. Previous success may depend on the commercial capabilities of TTO staff (O'Shea, Allen, and Chevalier 2005). Lockett and Wright (2005) assessed the relationship between the resources and capabilities of U.K. TTOs and the rate of start-up formation at their respective universities. They are able to distinguish empirically between a university's resource inputs and its routines and capabilities. Based on estimation of count regressions, the authors concluded that there is a positive correlation between start-up formation and the university's expenditure on intellectual property protection, the business development capabilities of TTOs, and the extent to which its royalty distribution formula favors faculty members. These findings imply that universities wishing to spawn numerous start-ups should devote greater attention to recruitment, training, and development of technology transfer officers with broad-based commercial skills.

More successful universities also develop vast social networks that help them generate more start-ups. Franklin, Wright, and Lockett (2001) analyze perceptions at U.K. universities regarding entrepreneurial start-ups that emerge from university technology transfer. The authors distinguish between academic and surrogate (external) entrepreneurs and "old" and "new" universities in the U.K. Old universities have well established research reputations, world-class scientists, and are typically receptive to entrepreneurial start-ups. New universities, on the other hand, tend to be weaker in academic research and less flexible with regard to entrepreneurial ventures. They find that the most significant barriers to the adoption of entrepreneurial-friendly policies are cultural and informational and that the universities generating the most start-ups (i.e., old universities) are those that have the most favorable policies regarding *surrogate* (external) entrepreneurs. The authors conclude that the best approach for universities that wish to launch successful technology transfer start-ups is a combination of academic and surrogate entrepreneurship. This would enable universities to simultaneously exploit the technical benefits of inventor involvement *and* the commercial know-how of surrogate entrepreneurs.

Wright et al. (2008) extend this analysis to mid-range universities across Europe in the context of all aspects of technology transfer, not just start-ups. Their evidence indicates that it may be difficult if not highly problematical for mid-range universities to attempt to emulate lessons from leading universities. The authors suggest that mid-range universities primarily need to focus on generating world-class research and critical mass in a few areas of expertise, rather than trying to be successful in many areas. Mid-range universities may need to develop a portfolio of university-industry linkages in terms of the scope of activities and the types of firms with which they interact, as well as with different types of intermediaries.

Incentives are also important. Lockett, Wright, and Franklin (2003) find that equity ownership was more widely distributed among the members of the spin-off company in the case of the more successful universities. Di Gregorio and Shane (2003) conclude that the ability of the university and inventor(s) to assume equity in a start-up, in lieu of licensing royalty fees, are significant determinants of the number of start-ups. They also find that a royalty distribution formula that is more favorable to faculty members reduces start-up formation, a finding that is confirmed by Markman et al. (2005). Di Gregorio and Shane (2003) attribute this result to the higher opportunity cost associated with launching a new firm, relative to licensing the technology to an existing firm.

Faculty quality, in terms of generating world class research of sufficient novelty, has also been shown to be positively associated with the start-up activity (Di Gregorio and Shane 2003; O'Shea, Allen, and Chevalier 2005).

Characteristics and Growth of Academic Start-Up Firms

Academic start-ups and spin-offs are heterogeneous. Wright et al. (2008) identified three distinct types of spin-offs. First, *venture capital backed type* spin-offs attract private venture capital very early in the start-up or founding phase. The venture capital is used both to further develop the technology and productize it in the market (Heirman and Clarysse 2005). Venture capital, however, implies that the company will have an "exit orientation." Typically, a venture capitalist wants to realize a trade sale. Therefore, the company will be structured at start-up in order to optimize the chances for such a trade sale. At start-up, this means that the venture should have a credible technology base and at least a roadmap to attract both a competent management team and further rounds of financing. Most universities are unlikely to have research groups with the world-leading and industrially applicable research capable of generating this kind of company.

Exactly the opposite of a VC-backed business model is the *life style* type of spin-off. These spin-offs start (and often remain) small in a market for products or services, often as consultancy services offered by academics around their research areas. Their value added as a population might be significant since they are large in number. Academics who already engage in significant amounts of industrial and contract research at universities usually find this type of company an extension of the activities they are already engaged in. As explained by Druihle and Garnsey (2004) this type of company offers the most familiar business model to the academic entrepreneur and therefore also has the highest chances of survival, which does not mean that it will create significant amounts of wealth, neither for the academic entrepreneur nor for the regional ecosystem in which it is embedded. From an academic entrepreneur point of view, the lifestyle company is based on knowledge rather than technology, uses a bootstrapping mode of finance, and tends to be fully managed by the academic or (part of) his former team of researchers. Confusion arises if these ventures are managed in the same way as venture capital backed spin-outs. A great deal of time and effort can be wasted attempting to "educate" the market to invest in a technology which clearly does not offer a sustainable competitive advantage.

Finally, the *prospector* category of companies is usually based on a niche technology, which needs to be further developed into a specific product before it can be sold. The investment to further develop the technology is too high to be financed by the academic entrepreneur, but too low to be attractive for a private venture capitalist. In this case, the academic entrepreneur in cooperation with the TTO manager face the challenge of attracting soft forms of venture capital, which are not too demanding in terms of business model pressure, but allow the academic entrepreneur to create a position on the market.

Spin-offs may adopt different types of growth strategy (Clarysse et al. 2007). A *product market strategy* is more likely to aim at achieving growth in terms of revenues to create a sustainable growing business. A *financial market strategy* places greater emphasis on creating value to enable the business to IPO, or be sold to a strategic partner. In this latter instance the strategy is to grow value which can be through building up the value of the science and technology in the business, even if no product sales are generated, or through a hybrid strategy of both building the value of the technology and generating sales, perhaps initially from consulting and services. The nature of the venture's sector and appropriability regime—for example, biotech versus ICT or engineering—may influence this choice. The choice of *commercialization strategy* will also be influenced by accessibility to the necessary complementary assets associated with each growth strategy (Gans and Stern 2003). *Product market* strategies imply a need to acquire human capital with commercialization expertise. *Financial market strategies* imply a need to access complementary human capital assets that can help develop the technology.

Using a hand-collected data set of 80 research-based spin-offs in five

European countries, Bruneel, Clarysse, and Wright (2009) show that research-based spin-off firms emphasizing a product strategy are positively associated with growth in revenues and that these kinds of firms that used a hybrid strategy achieved growth in both revenues and employment. However, firms pursuing a strategy to develop their technology do not grow fast in employment.

Despite considerable efforts to promote academic entrepreneurship, global evidence indicates that most public-research spin-offs are small and remain that way, even though these spin-offs are a fast-growing subpopulation of the entire population of young technology-based firms (Heirman and Clarysse 2007). For example, in France three-quarters are still in business six years after setup, but 80% of them have less than ten employees (Mustar 2001).

Using a sample of firms from Belgium, Clarysse, Wright, and Van de Velde (2011) find that a broad scope of technology has a positive and significant relationship with the growth of university spin-offs. As university spin-offs are often created to commercialize an invention based upon explorative activities within a university setting, they frequently start to develop their technology without probing into the market needs. Later on, they sometimes come to the conclusion that their technology is not well adjusted to the customer's needs or that the market is not yet ready (Vohora, Wright, and Lockett 2004). Therefore, they seem to benefit from maintaining a broader scope of technology that allows them to change market application if the first applications they pursue turn out to be a dead end (Siegel, Waldman, and Link 2003). A broad scope allows university spin-offs to explore in which market segment the technology has most value.

The importance of a broad scope of technology suggests an initial focus on building value rather than revenue streams. Scientific inventions that are narrower in scope may be less appropriate for the creation of university spin-offs, but instead may be more suitable for licensing. TTOs thus need to develop mechanisms and capabilities that enable them to sort scientific inventions into those that are suitable for licensing and those which can be developed as university spin-offs. These capabilities need to include both a research base of sufficient caliber to generate new technology and the skills to shape it into new products. The time scales likely to be involved in the development of products and services from university inventions emphasizes the need for longer-term support mechanisms with significant capabilities to create value (Clarysse et al.

2005). This in turn points to the importance of start-up capital, which emphasizes the need for policy support to ensure the availability of such capital for early-stage firms (Wright et al. 2006).

Colombo and Piva (2008) analyze the circumstances under which universities located in a geographical area contribute to the growth of academic start-ups. They examine the effects of a series of characteristics of local universities on the growth rates of academic start-ups and compare them with the effects of the same university characteristics on the growth of other (i.e., nonacademic) high-tech start-ups. The authors estimate an augmented Gibrat law panel data model using a longitudinal data set composed of 487 Italian firms observed from 1994 to 2003, of which 48 are academic start-ups. They find that universities influence the growth rates of local academic start-ups, while the effects on the growth rates of other start-ups are negligible. Importantly, the scientific quality of the research performed by universities has a positive effect on the growth rates of academic start-ups, yet the commercial orientation of research has a negative effect. This study demonstrates that universities producing high-quality scientific research have a beneficial impact on the growth of local high-tech start-ups, but only if these firms are able to detect, absorb, and use this knowledge. Thus the authors caution that a greater commercial orientation of university research that leads to a reduction of the knowledge available for absorption by these companies can be detrimental.

The exit of a spin-off through an IPO or strategic sale can be viewed as the final stage in the process of knowledge transfer involved in academic entrepreneurship. Of the large number of spin-offs created in recent years, few exit successfully through strategic sale or IPO. Library House data for the U.K. show that the vast majority of the spin-offs, more than 80%, are still active as independent companies. A minority of spin-offs realized an exit through an IPO (26) or via a trade sale (72). The IPO percentage is much higher than in a typical population of high tech firms where less than 5% make use of the IPO as an exit mechanism. Little is known, however, about which factors determine the probability of a trade sale, the time to trade sale, and the price paid at the moment of trade sale. This means that founder entrepreneurs often have to make managerial choices of which they cannot assess the impact.

Evidence suggests that few spin-offs fail (less than 10% according to Library House). Wennberg, Wiklund, and Wright (2011) compare the performance of university and corporate spin-offs. They show that corporate spin-offs (CSOs), especially those involving university graduates who had gone on to gain industrial experience, perform better than university spin-offs in terms of survival as well as growth.

Bonardo, Paleari, and Vismara (2010) analyze the dynamics of mergers and acquisitions for 131 science-based entrepreneurial firms (SBEFs) in Europe that went public during the period 1995–2003. They find that university affiliation enhanced attractiveness to other companies but was negatively related to the propensity to acquire. Bonardo, Paleari, and Vismara (2010) show that academic start-ups are more likely to engage in cross-border merger and acquisition activity. SBEFs may have very strong growth prospects as they seem more likely to have platform technologies than other high-tech start-ups (Clarysse, Wright, and Van de Velde, 2011); this may both make them attractive acquisition targets but also reduce their need for acquisitions.

Academic Entrepreneurs and Start-Ups

START-UPS AND RESEARCH PRODUCTIVITY. The seminal papers by Zucker and Darby and various collaborators explore the role of "star" scientists in the life sciences on the creation and location of new biotechnology firms in the U.S. and Japan. In Zucker, Darby, and Armstrong (2000), the authors assessed the impact of these university scientists on the research productivity of new U.S. biotech firms. Research productivity is measured using three proxies: number of patents granted, number of products in development, and number of products on the market. They find that ties between star scientists and firm scientists have a positive effect on these three dimensions of research productivity, as well as other aspects of firm performance and rates of entry in the U.S. biotechnology industry (Zucker, Darby, and Armstrong 1998; Zucker, Darby, and Brewer 1998).

An important consideration is whether engagement in venture creation comes at the opportunity cost of reduced publications. Ambos et al. (2008) show that star researchers are better able to be ambidextrous in terms of managing the heterogeneous objectives of teaching, research, and commercialization. Lowe and Gonzalez-Brambila (2007) analyze the research productivity of faculty members at 15 U.S. universities who formed a start-up company. These faculty members were more productive researchers than observationally equivalent colleagues *before* they established their firms. More importantly, the authors report that the research productivity of these academics did *not* decline in the aftermath of their entrepreneurial activity.

Based on comprehensive data from the University of Leuven in Belgium, Van Looy et al. (2004) find that engagement in entrepreneurial activity is associated with an increase in research output (publications), without affecting the nature of the publications involved. Nevertheless, some faculty who are not star researchers but who are frustrated by lack of academic recognition may also have research contracts with companies that generate possibilities for joint venture spin-offs.

In Zucker and Darby (2001), the authors examine detailed data on the outcomes of collaborations between "star" university scientists and biotechnology firms in Japan. Similar patterns emerge, in the sense that they find that such interactions substantially enhance the research productivity of Japanese firms, as measured by the rate of firm patenting, product innovation, and market introductions of new products. However, they also report an absence of geographically localized knowledge spillovers resulting from university technology transfer in Japan, in contrast to the U.S., where they found that such effects were strong. The authors attribute this result to the following interesting institutional difference between Japan and the U.S in university technology transfer. In the U.S., academic scientists typically work with firm scientists at the firm's laboratories. In Japan, firm scientists typically work in the academic scientist's laboratory. Thus, it is not surprising that the local economic development impact of university technology transfer appears to be lower in Japan than in the U.S.

Human Capital, Social Capital, and Start-Ups

Louis et al. (1989) analyze the propensity of life-science faculty to engage in various aspects of technology transfer, including commercialization. Their statistical sample consists of life scientists at the 50 research universities that received the most funding from the National Institutes of Health. The authors find that the most important determinant of involvement in technology commercialization was local group norms.

Yet, entrepreneurs at universities are different from other entrepreneurs in certain respects. Audretsch (2000) analyzes a data set on university life scientists, in order to estimate the determinants of the probability that they will establish a new biotechnology firm. Based on a hazard function analysis, including controls for the quality of the scientist's research, measures or regional activity in biotechnology, and a dummy for the career trajectory of the scientist, the author finds that university entrepreneurs tend to be older and more scientifically experienced. Other evidence also suggests that the academics developing high-tech ventures are typically key researchers in their fields (Vohora, Wright, and Lockett 2004).

However, the traditionally noncommercial university context poses major issues for the presence of the requisite human capital to enable academic start-ups to be successful. Rasmussen, Mosey, and Wright (2011) observe that the evolution of an academic spin-off beyond start-up requires the development of competencies or capabilities that involved the *creation* of new development paths that depart from existing practices in the academic context. The question then is what entrepreneurial competencies are needed for nascent spin-off ventures within a university context to reach the credibility threshold and, crucially, where do these competencies come from? Based on a longitudinal study involving 54 detailed interviews with academic entrepreneurs, TTOs, and other relevant parties in the U.K. and Norway, they identify three important competencies. First, they suggest that career academic entrepreneurs are distinctive in needing to develop the ability to attract new team members with industrial experience who can identify and interact with industrial partners to build an opportunity refinement competency. Second, they also show that career academic entrepreneurs are also more likely to need to evolve the credibility and entrepreneurial experience (leveraging competency) to enable interaction of the entrepreneurial team with external resource providers. Third, there is also a distinctive need to evolve the championing competency from the internal university context to include external champions. Given the general lack of industrial and entrepreneurial experience among academic entrepreneurs, gaining external champions residing within industrial partners or other resource providers may be particularly important in academic entrepreneurial ventures.

The social capital or networks of academic scientists who become entrepreneurs may be important influences on the performance of spin-offs. Mustar (1997) classified spin-offs depending on their cooperative arrangements with other public and/or private bodies and highlighted the relation between the breadth of the social network and the growth trajectory and attrition rate. Nicolaou and Birley (2003) recognized that different embeddedness of academics in a network of ties external or internal to the university may be associated with different growth trajectories.

However, differences in the human capital derived from the entrepreneurial experience of academic entrepreneurs may influence their ability to develop social capital that can address the barriers to venture development. Based on a longitudinal study of 24 academic entrepreneurs, supplemented with interviews with 20 technology transfer officers (TTOs) and heads of schools, Mosey and Wright (2007) report critical differences between the structure, content, and governance of social networks used to develop early stage ventures by three types of academic entrepreneurs with differing levels of entrepreneurship experience: nascent, novice, and habitual entrepreneurs. They suggest that habitual entrepreneurs (i.e., those with prior business ownership experience) have broader social networks and are more effective in developing network ties to gain equity finance and management knowledge. In contrast, less experienced entrepreneurs are likely to encounter structural holes between their scientific research networks and industry networks. This constrains their ability to recognize opportunities and gain credibility for their fledgling ventures. The authors also propose that while support initiatives, such as technology transfer offices and proof of concept funds, help attract industry partners to selected novice entrepreneurs, there appears to be no obvious substitute for business ownership experience to learn how to build relationships with experienced managers and potential equity investors. Experienced entrepreneurs were particularly critical of the ability of TTOs to provide useful contacts beyond a basic level. Interestingly, Mosey and Wright also find that development of social capital for nascent and novice entrepreneurs is influenced by the human capital related to the entrepreneurs' discipline base, with individuals from engineering and material sciences being more likely to build network ties outside their scientific research networks than those in biological sciences and pharmacy.

Similarly, Grimaldi and Grandi (2003) in a study of 40 Italian academic spin-offs find that the frequency of founding teams' interactions with external contacts is influenced by the frequency of interaction of the underlying research groups and by their scientific and technological excellence. The frequency of contacts with external agents positively influences the market attractiveness of the business idea (Grandi and Grimaldi 2005), as does the prior joint experience of the founders.

The growth of spin-offs may also require different types of skills and investment in human capital as encompassed in founding teams and subsequent boards of directors. Core founding teams in spin-offs appear to be unbalanced in terms of being highly experienced in research and development, but lacking in commercial functions and applied technological experience (Colombo and Piva 2008). Vanaelst et al. (2006) shed light on the evolution of entrepreneurial teams as the spin-off progresses from research to an independent venture. They show that some researchers who are actively involved in the first phase of the spin-off process exit and new members enter, especially those with commercial human capital. Exiting members make a career choice to stay with the university, preferring to continue their research but often having a part-time role providing the spin-off with continuing technological development.

As academic start-ups develop, they need to construct formalized boards, especially when the firm is created as a legal entity (Uhlaner, Wright, and Huse 2007). The boundaries of the founding team evolve into two new, overlapping teams—that is, the management team and the board. The board likely includes members of the founding team, the "privileged witnesses" (such as TTOs) who have been advising on the development of the firm, and new members such as venture capital representatives.

Bjørnåli and Gulbrandsen (2010) analyze eleven case studies of boards in academic start-ups. They explore board formation and changes in board composition occurring in Norwegian and U.S. spin-offs and find that the process of board formation is mainly driven by social networks of the founders. Although there are differences in the initial board compositions in Norwegian and U.S. firms, there is convergence over time in subsequent board changes, which are mainly driven by the social networks of the board chair. Additions of key board members are associated with the progress of an SBEF from one development stage to another.

Despite these welcome developments, empirical research so far does not adequately reflect the necessary life cycle in the development of boards as academic start-ups evolve. Filatotchev, Toms, and Wright (2006) and Zahra, Filatotchev, and Wright (2009) emphasize that boards serve different purposes as the venture develops. At start-up, where ownership and management typically overlap extensively, the firm may not need a monitoring system. Rather, a governance system which enhances resources and knowledge is likely to be more important. However, for academic start-ups to continue on their growth trajectory, there is a need to develop effective boards that can both sustain the ability to recognize and exploit entrepreneurial opportunities and protect external stakeholders', such as venture capitalists', interests. Zahra, Filatotchev, and Wright (2009) suggest that boards and absorptive capacity complement each other in enhancing entrepreneurship in academic start-ups that are moving beyond their initial stages of development.

Summary

The research on TTOs and start-up formation summarized in sections 3 and 4 underscores the importance of identifying the interests and incentives of those who manage the technology transfer process. These studies also highlight the importance of human capital (e.g., staffing of TTOs, "star scientists," and entrepreneurial teams) and the university culture (e.g., the role of department chairs, and entrepreneurs who are employed at these institutions) and group norms.

5. Challenges and Policy Recommendations

Our synthesis of the literature has identified some key challenges and some key "lessons learned" for university administrators, industry, and policy makers (e.g., regional or state authorities) in the U.S. (Siegel and Phan 2005) and E.U. (Wright et al. 2007) who want TTOs to be more "effective." In this context, "effectiveness" relates to generating additional revenue from commercializing university intellectual property and promoting the more rapid diffusion of knowledge from campus to the marketplace. In this section, we will address the stakeholders at the university level and their regional policy counterparts. Policy makers at the E.U. level are considered in section 6.

A key lesson learned from the academic literature is that universities should adopt a *strategic approach* to the commercialization of IP. First, they must consider a set of key formulation issues involving choices relating to institutional goals and priorities and consequent resource allocation. Given that universities are heterogeneous, in terms of their resource endowments and scientific base, these choices should reflect such configurations. Establishing priorities also relates to choices regarding technological emphasis for the generation of licensing opportunities, relating to *stage of development*. For instance, proof-of-concept technologies are likely to be more attractive than other technologies if the strategic objective is licensing for cash, since it is relatively easy to compute economic value under this scenario. Furthermore, such technologies can be codified for efficient arms-length transfer and they are more likely than other technologies to result in a commercial product, without substantial additional research expense.

Resource allocation decisions must also be driven by the increasing recognition that universities need to make strategic choices regarding the *mode of commercialization* they wish to emphasize—that is, licensing, start-ups, sponsored research and consulting, and other mechanisms of technology transfer that are focused more directly on stimulating economic and regional development, such as incubators and science parks.

University administrators, backed by regional policy makers, may also need to make a strategic choice regarding *technology field of emphasis*. Opportunities for technology commercialization and the propensity of faculty members to engage in technology transfer vary substantially across fields both between and within the life sciences and physical sciences (Wright, Birley, and Mosey 2004). There is also substantial variation in research quality across departments and college within a given university. If a university does not have a critical mass of research excellence or sufficient TTO expertise, that institution may need to establish a regional *collaboration*.

Universities also need to formulate *IP and patent strategies*. TTOs need to ensure that IP is well defined and protected before trying to raise commercial interests. This entails costs, in terms of recruiting sufficient expertise or paying for external advice. The ownership of IP also needs to be resolved. Thus, the IP and patent strategy should consider whether a technology is proprietary to the department, can be licensed on an exclusive base, or can be licensed on a nonexclusive basis. Relatedly, there is an emerging debate about universities becoming patent trolls through their retrospective attempts to generate income from inventions emanating from scientific activity. Rather than discovering new inventions or manufacturing the inventions covered by patents, patent trolls impose costs on the market by manipulating patents to extract financial gain for inventions they did not create. While universities enforcing their patents may have characteristics of trolls, given that they are not engaged in manufacturing a product, they are distinct from trolls in that they provide benefits from the new inventions they create (Lemley 2008).

Our review of the literature has shown the high opportunity cost of commercialization for academic entrepreneurs. Thus, there is a strong need for universities to adapt *promotion and tenure and remuneration systems* for academics so that commercialization activities are valued. The first major university to explicitly reward commercialization in promotion and tenure was Texas A&M in 2006. Based on a survey of North American institutions, Stevens, Johnson, and Sanberg (2011) reported that the following 16 universities in the U.S. and Canada now consider patents and commercialization in tenure and promotion decisions: Thompson Rivers University, University of Moncton, Northern Arizona University, Brigham Young University, Ohio University, University of North Carolina at Greensboro, George Mason University, University of Nebraska–UNeMed Corporation, Medical College of Wisconsin, Wake Forest University Health Sciences, Utah State, University of Texas Health Science Center Houston, Oregon State University, University of Saskatchewan, New York University, and University of Illinois at Urbana–Champaign. Since 2011, the University of Arizona and the University of Maryland have also instituted such policies.

Another potential impediment of "effective" technology commercialization is the ability of universities to attract and remunerate TTO personnel with the appropriate *skills* to support the commercialization strategy. For example, many universities have expanded the creation and development of start-ups. Traditionally, many TTO personnel have had a strong legal background, but are not well versed in the realm of entrepreneurship. This means that the requisite skills base for a given TTO employee extends beyond the need to protect intellectual property rights for the university to the need for opportunity recognition and exploitation, and other commercialization and entrepreneurial skills.

These challenges and evidence that few TTOs generate positive net income (Abrams, Leung, and Stevens 2009) have led to a questioning of their role at the university and in society. Some have suggested that legal ownership of inventions by universities is suboptimal, from an economic efficiency standpoint, and in terms of reducing the social benefits from the more rapid dissemination and commercialization of university-based research. According to this view, the TTO impedes commercialization and academic entrepreneurship, since it leads to delays in licensing, misalignment of incentives among parties, and delays in the flow of scientific information and the materials necessary for scientific progress (Kenney and Patton 2009).

An alternative approach is to vest ownership with the inventor, freeing them up to contract with whomever they see most able to assist in commercialization. However, as we have seen, unless this is supplemented with support policies that enable academic entrepreneurs to create value from the ventures they create, it is doubtful whether vesting ownership with the inventor will of itself lead to greater value creation. Another course of action is to adopt an open source strategy to make inventions publicly available or to be more selective in the use of exclusive licensing (Lemley 2008).

Some have argued that we need to integrate technology and knowledge transfer and TTOs into the curriculum and other aspects of the university (Martin 2012; Wright 2013). This involves going beyond direct technology and knowledge transfer to encompass indirect aspects. University education and research experience may lead indirectly to entrepreneurial actions, such as subsequent start-ups and corporate spin-offs, once graduates have gained industrial experience. As noted earlier, the performance effects of these ventures exceed those of university spin-offs. Further, a notable shift beyond spin-offs based on formal IP is associated with a growth in student start-ups based on new forms of technology that may be less demanding in terms of financing needs but which may require support to enable them to grow and create financial, economic, and social value. TTOs have a role to play in supporting entrepreneurial skills development and industry interactions for these student start-ups.

Some preliminary evidence from the U.S. and Europe suggests that business schools can play an important role in accelerating technology commercialization and entrepreneurship when they work effectively with a university TTO. One of the institutions in the vanguard of this movement is Johns Hopkins University. At Hopkins, the Carey Business School requires MBA students to take a "Discovery to Market" course, which involves a partnership with the Hopkins Tech Transfer Office to conduct a market analysis and commercialization plan for a university-based innovation. Other institutions in the U.S. where business schools work closely with the TTO include Oregon State University, RPI, the University at Albany, SUNY, the University of Montana, and the University of Wisconsin–Madison.

Notes

1. See Thursby, Jensen, and Thursby (2001) for an extensive description of this survey.

2. According to the Association of University Technology Managers (AUTM 2005), the number of start-up firms at U.S. universities rose from 35 in 1980 to 462 in 2004.

References

Abrams, I., G. Leung, and A. Stevens. 2009. "How are U.S. technology transfer offices tasked and motivated: Is it all about the money?" *Research Management Review* 17: 1–34.

Aghion, P., M. Dewatripont, and J. C. Stein. 2008. "Academic freedom, private-sector focus, and the process of innovation." *RAND Journal of Economics* 39 (3): 617–35.

Ambos, T., et al. 2008. "When does university research get commercialized? Creating ambidexterity in research institutions." *Journal of Management Studies* 45: 1424–47.

Arthurs, J. D., et al. 2008. "Managerial agents watching other agents: Multiple agency conflicts regarding underpricing in IPO firms." *Academy of Management Journal* 51: 277–94.

Association of University Technology Managers (AUTM). 2013. The AUTM Licensing Survey, Fiscal Year 2012. Norwalk, CT: AUTM, Inc.

Audretsch, D. 2000. "Is University Entrepreneurship Different?" Mimeo, Indiana University.

Belenzon, N., and M. Schankerman. 2009. "University Knowledge Transfer: Private Ownership, Incentives, and Local Development Objectives." *Journal of Law and Economics* 52 (1): 111–44.

Bercovitz, J., and M. Feldman. 2004. "Academic Entrepreneurs: Social Learning and Participation in University Technology Transfer." Mimeo, University of Toronto.

Bercovitz, J., and M. Feldman. 2008. "Academic Entrepreneurs: Organizational Change at the Individual Level." *Organization Science* 19: 69–89.

Bercovitz, J., et al. 2001. "Organizational Structure as Determinants of Academic Patent and Licensing Behavior: An Exploratory Study of Duke, Johns Hopkins, and Pennsylvania State Universities." *Journal of Technology Transfer* 26: 21–35.

Bjørnåli, E., and M. Gulbrandsen. 2010. "Exploring board formation and evolution of board composition in academic spin-offs. *Journal of Technology Transfer* 35: 92–112.

Bonardo, D., S. Paleari, and S. Vismara. 2010. "The M&A Dynamics of European Science Based Entrepreneurial Firms." *Journal of Technology Transfer* 35: 141–80.

Bruneel, J., B. Clarysse, and M. Wright. 2009. "Linking Entrepreneurial Strategy and Firm Growth." Working Papers of Faculty of Economics and Business Administration, Ghent University, Belgium, Faculty of Economics and Business Administration.

Bruton, G., et al. 2010. "Governance, ownership structure and performance of IPO firms: the impact of different types of private equity investors and institutional environments." *Strategic Management Journal* 31(5): 491–509.

Carlsson, B., and A. Fridh. 2002. "Technology Transfer in United States Universities: A Survey and Statistical Analysis." *Journal of Evolutionary Economics* 12: 199–232.

Chandler, A. 1962. *Strategy and Structure: Chapters in the History of the American Industrial Enterprise*. Cambridge, MA: MIT Press.

Chapple, W., et al. 2005. "Assessing the Relative Performance of University Technology Transfer Offices in the U.K.: Parametric and Non-Parametric Evidence." *Research Policy* 34 (3): 369–84.

Clarke, B. R. 1998. *Creating Entrepreneurial Universities; Organizational Pathways of Transformation*. New York: IAU Press.

Clarysse, B., et al. 2005. "Spinning out new ventures: A typology of incubation strategies from European research institutions." *Journal of Business Venturing* 20 (2): 183–216.

Clarysse, B., et al. 2007. "Academic Spin-offs, Formal Technology Transfer and Capital Raising." *Industrial and Corporate Change* 16: 609–40.

Clarysse, B., M. Wright and E. Van de Velde, 2011. "Entrepreneurial Origin, Technology Endowments and the Growth of Spin-off Companies." *Journal of Management Studies* 48: 1420–42.

Colombo, M. G., and E. Piva. 2008. "Firms' genetic characteristics, competence enlarging strategies, and performance: a comparison of academic and non-academic start-ups." Politecnico di Milano, Department of Management, Economics and Industrial Engineering, working paper.

Debackere, K., and R. Veugelers. 2005. "The Role of Academic Technology Transfer Organizations in Improving Industry Science Links." *Research Policy* 34 (3): 321–42.

Degroof, J. J., and E. B. Roberts. 2004. "Overcoming Weak Entrepreneurial Infrastructure for Academic Spin-off Ventures." *Journal of Technology Transfer* 29 (3–4): 327–57.

Di Gregorio, D., and S. Shane. 2003. "Why Do Some Universities Generate More Start-Ups than Others?" *Research Policy* 32: 209–27.

Druilhe, C., and E. Garnsey. 2004. "Do Academic Spin-Outs Differ and Does It Matter?" *Journal of Technology Transfer* 29 (3–4): 269–285.

Filatotchev, I., S. Toms, and M. Wright. 2006. "The firm's strategic dynamics and corporate governance life-cycle." *International Journal of Managerial Finance* 2 (4): 256–79.

Fini, R., R. Grimaldi, and M. Sobrero. 2009. "Factors Fostering Academics to

Start-up New Ventures: An Assessment of Italian Founders' Incentives." *Journal of Technology Transfer* 34: 380–402.

Franklin, S., M. Wright, and A. Lockett. 2001. "Academic and surrogate entrepreneurs in university spin-out companies." *Journal of Technology Transfer* 26 (1–2): 127–41.

Friedman, J., and J. Silberman. 2003. "University technology transfer: Do incentives, management, and location matter?" *Journal of Technology Transfer* 28 (1): 81–85.

Gans, J. S. and S. Stern. 2003. "The product market and the market for 'ideas': commercialization strategies for technology entrepreneurs." *Research Policy* 32 (2): 333–50.

Grandi, A., and R. Grimaldi. 2005. "Academics' organizational characteristics and the generation of successful business ideas." *Journal of Business Venturing* 20 (6): 821–45.

Grimaldi, R., and A. Grandi. 2003. "Exploring the networking characteristics of new venture founding teams: a study of Italian academic spin-offs." *Small Business Economics* 21 (4): 329–41.

Grimaldi, R., et al. 2011. "30 years after Bayh-Dole: Reassessing academic entrepreneurship." *Research Policy* 40 (8): 1045–57.

Heirman, A., and B. Clarysse. 2005. "The imprinting effect of initial resources and market strategy on the early growth path of start-ups." Working Papers of Faculty of Economics and Business Administration, Ghent University, Belgium, 05/310, Faculty of Economics and Business Administration.

Heirman A., and B. Clarysse B. 2007. "Which tangible and intangible assets matter for innovation speed in start-ups?" *Journal of Product Innovation Management* 24: 303–15.

Hellmann, T. 2007. "The Role of Patents for Bridging the Science to Market Gap." Mimeo. *Journal of Economic Behavior & Organization* 63: 624–47.

Hertzfeld, H. R., A. N. Link, and N. S. Vonortas. 2006. "Intellectual property protection mechanisms in research partnerships." *Research Policy* 35: 825–39.

Hoppe, H., and E. Ozdenoren. 2005. "Intermediation in Innovation." *International Journal of Industrial Organization* 23: 483–503.

Jensen, R., and M. Thursby. 2001. "Proofs and Prototypes for Sale: The Licensing of University Inventions." *American Economic Review* 91 (1): 240–59.

Jensen, R., J. G. Thursby, and M. C. Thursby. 2003. "The Disclosure and Licensing of University Inventions: The Best We Can Do with the S**t We Get to Work With." *International Journal of Industrial Organization* 21 (9): 1271–1300.

Kenney, M., and D. Patton. 2009. "Reconsidering the Bayh-Dole Act and the Current University Invention Ownership Model." *Research Policy* 38: 1407–22.

Lacetera, N. 2009. "Academic Entrepreneurship." *Managerial and Decision Economics* 30 (7): 443–64.

Lach, S. and M. Schankerman. 2004. "Royalty Sharing and Technology Licensing in Universities." *Journal of the European Economic Association* 2 (2–3): 252–64.

Lemley, M. 2008. "Are Universities Patent Trolls?" *Fordham Intellectual Property, Media and Entertainment Law Journal* 18: 611; Stanford Public Law Working Paper No. 980776. Available at SSRN: http://ssrn.com/abstract=980776.

Link, A. N., and D. S. Siegel. 2005. "Generating Science-Based Growth: An Econometric Analysis of the Impact of Organizational Incentives on University-Industry Technology Transfer." *European Journal of Finance* 11 (3): 169–82.

Lockett, A., and M. Wright. 2005. "Resources, Capabilities, Risk Capital and the Creation of University Spin-Out Companies." *Research Policy* 34 (7): 1043–57.

Lockett, A., M. Wright, and S. Franklin. 2003. "Technology Transfer and Universities' Spin-out Strategies." *Small Business Economics* 20: 185–201.

Lockett, A., et al. 2005. "The Creation of Spin-off Firms at Public Research Institutions: Managerial and Policy Implications." *Research Policy* 34 (7): 981–93.

Louis, K., et al. 1989. "Entrepreneurs in Academe: An Exploration of Behaviors Among Life Scientists." *Administrative Science Quarterly* 34: 110–31.

Louis, K., et al. 2001. "Entrepreneurship, Secrecy, and Productivity: A Comparison of Clinical and Non-Clinical Life Sciences Faculty." *Journal of Technology Transfer* 26 (3): 233–45.

Lowe, R. A., and C. Gonzalez-Brambila. 2007. "Faculty Entrepreneurs and Research Productivity." *Journal of Technology Transfer* 32: 173–94.

Macho-Stadler, I., X. Martinez-Giralt, and D. Pérez-Castrillo. 1996. "The Role of Information in Licensing Contract Design." *Research Policy* 25: 43–57.

Macho-Stadler, I., D. Pérez-Castrillo, and R. Veugelers. 2007. "Licensing of University Innovations: The Case of a Technology Transfer Office." *International Journal of Industrial Organization* 25 (3): 483–510.

Macho-Stadler, I., D. Pérez-Castrillo, and R. Veugelers. 2008. "Designing Contracts for University Spin-offs." *Journal of Economics and Management Strategy* 17 (1): 185–218.

Markman, G., et al. 2005. "Entrepreneurship and University-Based Technology Transfer." *Journal of Business Venturing* 20 (2): 241–63.

Markman, G., P. Gianiodis, and P. Phan. 2008. "Full-time faculty or part-time entrepreneurs." *IEEE Transactions on Engineering Management* 55: 29–36.

Martin, B. 2012. "Are universities and university research under threat? Towards an evolutionary model of university speciation." *Cambridge Journal of Economics* 36: 543–65.

Mosey, S., and M. Wright. 2007. "From human capital to social capital: A longitudinal study of technology-based academic entrepreneurs." *Entrepreneurship Theory and Practice* 31, 909–36.

Mustar, P. 1997. "Spin-off enterprises: How French academies create hi-tech companies: The condition for success or failure." *Science and Public Policy* 24 (1): 37–43.

Mustar, P. 2001. "Spin-offs from public research: Trends and outlook." *STI Review* 26: 165–72.

Mustar, P., et al. 2006. "Conceptualising the heterogeneity of research-based spin-offs: A multi-dimensional taxonomy." *Research Policy* 35 (2): 289.

Nicolaou, N., and S. Birley. 2003. "Social networks in organizational emergence: The university spinout phenomenon." *Management Science* 49 (12): 1702–25.

Organisation for Economic Co-operation and Development (OECD). 2001. *Benchmarking Industry-Science Relationships, Science, Technology and Industry Outlook*. Paris: OECD Publications.

O'Shea, R., T. Allen, and A. Chevalier. 2005. "Entrepreneurial Orientation, Technology Transfer, and Spin-off Performance of U.S. Universities." *Research Policy*, 34 (7): 994–1009.

Perez, M., and A. M. Sanchez. 2003. "The development of university spin-offs: Early dynamics of technology transfer and networking." *Technovation* 23 (10): 823–31.

Phan, P., and D. S. Siegel. 2006. "The effectiveness of university technology transfer: Lessons learned, managerial and policy implications, and the road forward." *Foundations and Trends in Entrepreneurship* 2 (2): 77–144.

Powell, W. W., and J. Owen-Smith. 1998. "Universities and the market for intellectual property in the life sciences." *Journal of Policy Analysis and Management* 17 (2): 253–77.

Powers, J. B., and P. McDougall. 2005a. "University start-up formation and technology licensing with firms that go public: A resource-based view of academic entrepreneurship." *Journal of Business Venturing* 20 (3): 291–311.

Powers, J. B., and P. McDougall. 2005b. "Policy orientation effects on performance with licensing to start-ups and small companies." *Research Policy* 34 (7): 1028–42.

Poyago-Theotoky, J., J. Beath, and D. S. Siegel. 2002. "Universities and fundamental research: Reflections on the growth of university-industry partnerships." *Oxford Review of Economic Policy* 18 (1): 10–21.

Rasmussen, E., S. Mosey, and M. Wright. 2011. "The evolution of entrepreneurial competencies: A longitudinal study of university spin-off venture emergence." *Journal of Management Studies* 48: 1314–45.

Rasmussen, E., S. Mosey, and M. Wright. 2014. "The influence of university departments on the evolution of entrepreneurial competencies in spin-off ventures." *Research Policy* 43:92–106.

Roberts, E. 1991. *Entrepreneurs in High Technology, Lessons from MIT and Beyond*. Oxford: Oxford University Press.

Roberts, E., and D. E. Malone. 1996. "Policies and structures for spinning off new companies from research and development organizations." *R&D Management* 26: 17–48.

Rogers, E. M., Y. Yin, and J. Hoffmann. 2000. "Assessing the effectiveness of technology transfer offices at U.S. research universities." *Journal of the Association of University Technology Managers* 12: 47–80.

Rothaermel, F. T., S. Agung, and L. Jian. 2007. "University entrepreneurship: A taxonomy of the literature." *Industrial and Corporate Change* 16 (4): 691–791.

Shrader, R., and D. S. Siegel. 2007. "Assessing the relationship between human capital and firm performance: evidence from technology-based new ventures." *Entrepreneurship Theory and Practice* 31: 893–908.

Siegel, D. S., and P. Phan. 2005. "Analyzing the effectiveness of university technology transfer: Implications for entrepreneurship education." In *Advances in the Study of Entrepreneurship, Innovation, and Economic Growth*, edited by Gary Liebcap. Amsterdam: Elsevier Science/JAI Press, vol. 16, 1–38.

Siegel, D. S., D. A. Waldman, and A. N. Link. 2003. "Assessing the impact of organizational practices on the productivity of university technology transfer offices: An exploratory study." *Research Policy* 32 (1): 27–48.

Siegel, D. S., P. Westhead, and M. Wright. 2003. "Assessing the impact of science parks on the research productivity of firms: Exploratory evidence from the United Kingdom." *International Journal of Industrial Organization* 21 (9): 1357–69.

Siegel, D. S., et al. 2003. "Commercial knowledge transfers from universities to firms: Improving the effectiveness of university-industry collaboration." *Journal of High Technology Management Research* 14: 111–33.

Siegel, D. S., et al. 2004. "Toward a model of the effective transfer of scientific knowledge from academicians to practitioners: Qualitative evidence from the commercialization of university technologies." *Journal of Engineering and Technology Management* 21 (1–2): 115–42.

Siegel, D., et al. 2008. "Assessing the Relative Performance of University Technology Transfer in the U.S. and U.K.: A Stochastic Distance Function Approach." *Economics of Innovation and New Technology* 17: 717–29.

Stevens, A. J., G. A. Johnson, and P. R. Sanberg. 2011. "The role of patents and commercialization in the tenure and promotion process." *Technology and Innovation* 13: 241–48.

Thursby, J. G., R. Jensen, and M. C. Thursby. 2001. "Objectives, characteristics and outcomes of university licensing: A survey of major U.S. universities." *Journal of Technology Transfer* 26: 59–72.

Thursby, J. G., and S. Kemp. 2002. "Growth and productive efficiency of university intellectual property licensing." *Research Policy* 31: 109–24.

Thursby, J. G., and M. C. Thursby. 2002. "Who is selling the ivory tower? Sources of Growth in University Licensing," *Management Science* 48: 90–104.

Thursby, J. G., and M. C. Thursby. 2004. "Are faculty critical? Their role in university licensing." *Contemporary Economic Policy* 22 (2): 162–78.

Thursby, J. G., and M. C. Thursby. 2007. "University licensing." *Oxford Review of Economic Policy* 23(4): 620–39.

Uhlaner, L., M. Wright, and M. Huse. 2007. "Private firms and corporate governance: An integrated economic and management perspective." *Small Business Economics* 29: 225–41.

Vanaelst, I., et al. 2006. "Entrepreneurial team development in academic spinouts: An examination of team heterogeneity." *Entrepreneurship Theory and Practice* 30: 249–72.

Van Looy, B., et al. 2004. "Combining entrepreneurial and scientific performance in academia: Towards a compound and reciprocal Matthew-effect?" *Research Policy* 33: 425–41.

Vohora, A., M. Wright, and A. Lockett. 2004. "Critical junctures in the development of university high-tech spinout companies." *Research Policy* 33 (1): 147–75.

Wennberg, K., J. Wiklund, and M. Wright. 2011. "The effectiveness of university knowledge spillovers: Performance differences between university spin-offs and corporate spin-offs." *Research Policy* 40 (8): 1128–43.

Williamson, O. E. 1975. *Markets and hierarchies: Analysis and antitrust implications.* New York: Free Press.

Wright, M. 2014. "Academic entrepreneurship, technology transfer and society: Where next?" *Journal of Technology Transfer.*

Wright, M., S. Birley, and S. Mosey. 2004. "Entrepreneurship and university technology transfer." *Journal of Technology Transfer* 29 (3/4): 235–46.

Wright, M., et al. 2006. "University spin-out companies and venture capital." *Research Policy* 35 (4): 481–501.

Wright, M., et al. 2007. *Academic entrepreneurship in Europe.* Cheltenham, U.K.: Edward Elgar.

Wright, M., et al. 2008. "Mid-range universities' in Europe linkages with industry: Knowledge types and the role of intermediaries." *Research Policy* 37: 1205–23.

Wright, M., et al. 2009. "Academic entrepreneurship and business bchools." *Journal of Technology Transfer* 34: 560–87.

Zahra, S., I. Filatotchev, and M. Wright. 2009. "How do threshold firms sustain corporate entrepreneurship? The role of boards of directors and knowledge. *Journal of Business Venturing* 24: 248–60.

Zucker, L. G., and M. R. Darby. 2001. "Capturing technological opportunity via Japan's star scientists: Evidence from Japanese firms' biotech patents and products." *Journal of Technology Transfer* 26 (1–2): 37–58.

Zucker, L. G., M. R. Darby, and J. Armstrong. 1998. "Geographically localized knowledge: Spillovers or markets?," *Economic Inquiry* 36 (1): 65–86.

Zucker, L. G., M. R. Darby, and J. Armstrong, J. 2000. "University science, venture capital, and the performance of U.S. biotechnology firms." Mimeo, UCLA.

Zucker, L. G., M. R. Darby, and M. B. Brewer. 1998. "Intellectual Human Capital and the Birth of U.S. Biotechnology Enterprises." *American Economic Review* 88 (1): 290–306.

CHAPTER TWO

Open Science and Open Innovation

Sourcing Knowledge from Universities

Markus Perkmann and Joel West

Introduction

Open innovation means that firms source knowledge and technology from outside their boundaries in their efforts to innovate (Chesbrough 2003). Combining external search with internal R&D capabilities provides firms with a larger spectrum of technological options and allows them to dedicate relatively more resources to R&D areas they consider their core advantages.

Firms have engaged in open innovation since the inception of formal R&D. In this chapter, we focus on a specific mode of open innovation, collaboration with universities.[1] The recent focus on developments following the U.S. Bayh-Dole Act of 1980 has perhaps obscured the fact that firms' engagement with universities has a long history and dates back to the earliest systematic efforts by firms to conduct R&D (Mowery 2009). For instance, dense network relationships between university laboratories and chemical firms enabled the rise of the German synthetic dye industry (Murmann 2003) and the creation of firms in the Bay area (Lenoir 1997; Kenney 1986). As early as the 1930s, MIT developed policies for faculty consulting and licensing its patents to industry (Etzkowitz 2002).

From the perspective of the firm, collaborating with universities differs from collaborating with commercial entities in one fundamental respect. Universities are the locus where most publicly funded research is conducted, and most public research forms part of "open science" (Dasgupta and David 1994). Open science means that research is conducted with the objective of publishing results. Even though the conduct of open science is not necessarily in conflict with commercial exploitation (Murray 2002), in practice there are potential points of friction that can affect collaboration between firms and universities (Slaughter and Leslie 1997; Krimsky 2003; Stuart and Ding 2006).

Universities' primary activities are to educate students and to conduct "open science" for publication, so serving commercial clients is conducted side by side with these activities. There are two structural aspects of university life that may represent barriers for successful interactions with firms. First, academics' choices of research topics will often reflect their intention to seek novel scientific contributions, rather than maximizing the commercial usefulness of a research program. Novel scientific contributions provide priority in publishing which in turn forms the basis for accumulating reputation and status within the professional community of academia (Merton 1973).

Second, firms' efforts to protect and appropriate knowledge to create commercial advantage may be at odds with the logic of open science. Intellectual property protection may obstruct or decelerate the free flow of ideas necessary for open science (David 2004; Murray and Stern 2007). Furthermore, intellectual property rights creates transaction costs that may limit access to university knowledge or research materials (Murray 2010), while the pre-patent secretiveness of university researchers may reduce the flow of knowledge from universities to other public and private researchers (Fabrizio 2006). Within the university, the income disparities and envy arising from these commercial ties can fracture the broader university cultural norms (Argyres and Liebeskind 1998).

However, despite the cultural and structural differences between open science and commercial science, firms have much to gain from the ongoing production of publicly funded research. Universities play a crucial role in generating and disseminating new knowledge into the broader economy (Henderson, Jaffe, and Trajtenberg 1998; Cohen and Levinthal 1990). Universities are an important external source of innovations for firms, particularly in science-intensive sectors such as chemicals and pharmaceutical (Mansfield 1991; Cohen, Nelson, and Walsh 2002; Laursen and Salter 2004; Hanel and St-Pierre 2006). In some cases, university innovations and their antecedents are accessed by firms via spillovers through the open literature and flows of researchers and students (Salter and Martin 2001; Mohnen and Hoareau 2003). Mansfield's (1991) study suggested that approximately a tenth of corporate innovations are based directly on academic research.

Rather than such spillovers, in this chapter we focus on knowledge transfer and creation through the direct and formal interaction of firms and universities. Ways for firms to directly access university knowledge assumes different forms, including licensing, research collaborations, contract research, sponsored research, and faculty consulting (Link, Siegel, and Bozeman 2007; D'Este and Patel 2007; Perkmann and Walsh 2007). The diversity of channels through which knowledge and technology flow between firms and universities testifies that firms work with universities not just to access "novel" technological knowledge but also for supporting and completing ongoing development activities (MacPherson 2002; Cohen et al. 2002; Carayol 2003; Lee 2000).

While universities are a source of knowledge and innovations for firms, particularly in some sectors, extant research on "open innovation" has paid relatively scant attention to university-industry relations (but see Fabrizio 2006; Perkmann and Walsh 2007). We therefore review the broader literature on firm-university interactions from the firm's viewpoint, identify how it informs our knowledge of this specific variant of open innovation, and then suggest opportunities for further research that integrates these two literatures.

Below, we provide an overview of three main modes of direct interaction between firms and universities: IP licensing, research services, and research partnerships.[2] We outline the main characteristics of each mode, its relative importance for firms as well as benefits and challenges. While licensing remains an important mode in which public research finds its way into firms' laboratories, we highlight the significant role of relationship-based modes of interaction—research services and research partnerships—between firms and universities. While some of these relationship-based interactions enable appropriation via intellectual property rights, others are more aligned with the norms of open science and create benefits for firms by generating basic knowledge, creating skills, and enabling follow-on innovation. We provide a tabular summary of all referenced empirical articles on sourcing knowledge from universities in table 2.2 at the end of the chapter.

Three Ways of Sourcing Innovation from Universities

Links between firms and universities can be seen as part of the more general open innovation picture. The generic external benefits of universities, such as educating cohorts of graduates, generating scientific knowledge and instrumentation, have long been recognized as an important source of industrial innovation (Pavitt 1991; Salter and Martin 2001). In the wake of the Bayh-Dole Act and related policy initiatives in other countries, universities have made efforts to more extensively engage in industrial innovation, resulting in an increasing patenting propensity by universities (Nelson 2001), growing university revenues from licensing (Thursby, Jensen, and Thursby 2001), increasing numbers of university researchers engaging in academic entrepreneurship (Shane 2005), the diffusion of technology transfer offices, industry collaboration support offices and science parks (Siegel, Waldman, and Link 2003), and a growing number of university spin-out companies (Lockett and Wright 2005).

One way in which firms access university technologies is by *licensing intellectual property* owned by universities. Licensing refers to contractual agreements according to which firms acquire the right to use university-generated intellectual property against a fee, commonly via licensing (Phan and Siegel 2006). While firms usually prefer exclusive rights to such IP, some IP may be licensed on a nonexclusive basis to multiple firms. For instance, Stanford and the University of California licensed their recombinant DNA technology to several hundred firms (Bera 2009). Licensing represents a common "outside-in" path for firms to gain access to existing technologies relatively rapidly (Chesbrough 2003; Enkel, Gassmann, and Chesbrough 2009). Yet in-licensing of intellectual property represents only one way in which firms access university-generated knowledge (Cohen, Nelson, and Walsh 2002; Arundel and Geuna 2004; Faulkner and Senker 1994). Indeed, the transfer of codified knowledge, via patenting and licensing, appears to be only moderately important compared to other modes of interaction (Agrawal and Henderson 2002; Schartinger et al. 2002; Cohen, Nelson, and Walsh 2002). Surveys suggests that U.S. R&D executives place the highest value on contract research, consulting, and cooperative research, while licensing is judged as less relevant (Roessner 1993; Cohen, Nelson, and Walsh 2002). In our discussion below, we describe two main mechanisms that do not primarily rely on IP transfer: research services and research partnerships. These two modes correspond to the "coupled path" of open innovation where firms and external partners combine their own technology to jointly produce (and often commercialize) technology of value to the firm (Enkel, Gassmann, and Chesbrough 2009).[3]

Research services are paid-for services performed by university researchers for external clients (i.e., consulting and contract research) (Perkmann and Walsh 2007). Consulting refers to activities in which an academic provides advice and expertise, usually for personal income, while contract research usually involves the use of university equipment whereby the generated proceeds benefit the university or research group. Both forms of research services typically include explicit IP license agreements but—in contrast to the IP transfer case—involve developing new IP for the firm's custom specifications rather than licensing of IP already developed by university employees. Researchers may also offer their scientific expertise by serving on boards of directors or scientific advisory boards. Studies of the former suggest that academic scientists who are directors help management allocate investments in R&D and other technological projects (Hülsbeck and Lehmann 2012; White et al. 2013). Academics on scientific advisory boards provide ties for future university sourcing, while legitimating firms (particularly under high technological uncertainty) with external stakeholders (Stuart, Ozdemir, and Ding 2007; Chok 2009).

Research partnerships are formal collaborative arrangements with the objective to cooperate on research and development activities (Hall, Link, and Scott 2001). Such relationships are often referred to as "collaborative research," "joint research," or "research joint ventures." They can range from small-scale, temporary projects to larger, long-term university-industry research centers to permanent, large-scale consortia with hundreds of industrial members. Such partnerships differ from research services in that they include either research contributions by the firms, alignment to the university's nonmonetary goals (such as publication), or both.

The three modes of sourcing university knowledge differ with respect to the incentives they provide to university researchers. While licensing generates an economic rent that is usually shared between inventors and their institution, research services and research partnerships also provide non-pecuniary benefits to the researcher. The modes also differ in the degree of control they provide to the firm, the access and influence over the direction of research, and the risks and costs of sourcing. Below,

TABLE 2.1. **Three modes of sourcing university innovation.**

	IP transfer	Research services	Research partnerships
Approach	Acquire existing ideas needed for product or process innovation	Contract to develop new ideas in well-defined areas	Collaborate to discover new ideas
Firm goals	Rapidly obtain technology developed by public science	Outsource R&D projects instead of developing internal capabilities	Leverage the respective strengths of internal and external researchers, and access public science
University goals	Income from academic inventions; economic impact; service to faculty	Income from leveraging existing expertise; access to ideas, materials, and data	Shared scientific advance; income from public and private sources
Required firm capabilities	Search and external monitoring	Purchasing, contract management	Relationship management, learning
Application areas	Sectors with high relevance of intellectual property	Mature sectors with predictable innovation trajectories	Newly emerging technology areas and industrial sectors, science-intensive sectors

we discuss the three modes in more depth, outlining in detail their purposes, the circumstances in which they are used, and the capabilities firms require to engage in them. We provide a tabular summary in table 2.1.

Licensing

Licensing an invention patented or otherwise protected by a university enables firms to initiate new pathways of technology development, as when biotechnology companies license a novel molecular entity discovered in a university laboratory. A number of countries have designed policies that provide universities with incentives to protect their intellectual property and make it accessible to external organizations (Mowery and Sampat 2005). Many universities have established technology transfer offices with a mandate to identify commercializable inventions, protect the university's intellectual property rights, and initiate licensing deals with commercial buyers (Owen-Smith and Powell 2001; Debackere and Veugelers 2005; Siegel, Waldman, and Link 2003). Licensing is relevant particularly in industries where patented intellectual property plays an important role, such as in biotechnology and information technology (Niosi 2006; Brusoni, Marsili, and Salter 2005).

In terms of the type of knowledge being sourced, most licensing deals with universities involve early-stage technologies (Thursby and Thursby 2004; Colyvas et al. 2002), although on some occasions licenses may provide complements and improvements for down-stream technology development projects pursued by firms. At an early stage, a new technology tends to be characterized by high technical or commercial uncertainty (Jensen and Thursby 2001). Such uncertainties may not be resolved until the licensee makes a considerable investment to commercialize the technology, further increasing the risk borne by the firm.

In order to minimize such risk, licensing arrangements are often supplemented by a partnering arrangement involving ongoing involvement of the inventor (Agrawal 2006; Thursby and Thursby 2004). The underlying reason for such inventor engagement is—particularly in the case of novel technologies—that a considerable part of related expertise is usually not available in codified form, and hence the inventor can claim exclusivity (Zucker, Darby, and Armstrong 2002). Engaging the inventor through partnership therefore represents a mechanism for firms to "capture" rare knowledge that is too expensive to codify when its perceived value is low but is quickly eroded by competition when its value is high (Zucker, Darby, and Armstrong 2002). This enables a firm to accelerate technology development and hence enjoy first-mover advantages before the expertise diffuses via codification (Agrawal 2006).

Apart from inventor involvement, development risks and other principal-agent problems are also commonly mitigated by sharing the risk with the university, for instance by linking payments to the eventual commercial proceeds from the technology (Jensen and Thursby 2001). This may be done through royalties on products incorporating the specific invention.

A significant question for potential corporate licensees is how valuable university technologies are compared to other patents. The evidence is mixed but indicates that university technology has value particularly in science-intensive sectors. A study of biotechnology firms, for instance, indicates that firms with university linkages (which include licenses) were more innovative than those that did not (George, Zahra, and Wood 2002). In a comparison of academic and industry patents, Henderson, Jaffe, and Trajtenberg (1998) found that university patents before 1980 were of higher quality (i.e., more highly cited) than the average U.S. patent. However, Sapsalis, Van Pottelsberghe De La Potterie, and Navon (2006) found that university biotech patents were comparable in quality to industry patents in the same families filed in Belgium from 1995 to 1999.

In summary, engaging in IP transfer enables a firm to achieve relatively rapid access to an existing technology that is unavailable internally. Accessing externally generated technology, even if it is embryonic, may allow a firm to embark on a new product development pathway or provide complements to existing pathways.

Firms require specific capabilities to take advantage of this specific way of accessing university knowledge. First, as potentially useful ideas are likely to be widely dispersed, firms require adequate search and monitoring capabilities. The ability to span organizational as well as technology boundaries allows firms to overcome the lock-in effects of local search, resulting in higher degrees of innovativeness (Rosenkopf and Nerkar 2001). To search widely (across a number of external organizations) and deeply (intensively across their collaborating organizations) has a positive impact on firm innovativeness as long as a firm's resources are adequate to cope with the number of contacts initiated (Laursen and Salter 2006). In reality, firms appear to rely on personal contacts to a significant degree when searching for technology to license from universities, indicating a risk of lock in (Thursby and Thursby 2001).

Second, organizational capabilities are needed to support technology licensing. This includes the establishment of specialized units for intellectual property management that might act as a profit center across corporations (Arora, Fosfuri, and Gambardella 2001; Rivette and Kline 2000; Siegel, Waldman, and Link 2003).

Finally, because of its specific characteristics, licensing works better in some circumstances than others. Of particular relevance are the appropriability conditions under which it occurs. Acquiring technology in this way works best when there is a significant gap between replication and imitation costs (Teece 1986). In other words, if the underlying technology can be clearly codified and articulated (Winter 1987) and intellectual property rights can be well defined and protected, licensing will work. Such conditions prevail for instance in the chemical and pharmaceutical industries as well as high-tech sectors such as software or semiconductors (Levin et al. 1987).

Research Services

A second way of accessing university knowledge is via research services, notably via contract research or consulting (Perkmann and Walsh 2007). Such services usually involve an explicit and formal assignment of IP rights from the university (or university researcher) to the firm. A crucial difference between research services and licensing is that the latter involves an invention that already exists while in the case of research services, the emphasis is on collaboration that may or may not be structured around existing IP.

Contract research means a firm commissions a university to conduct a specified research project for the payment of a fee (Cassiman, Di Guardo, and Valentini 2010). In many cases, these contracts stipulate confidentiality and hence it may not be possible for the collaborating academics to openly publish the research results. Similarly, they may also specify for any intellectual property arising from fully costed research contracts to be assigned to the commissioning party.

An alternative to a research contract with a university is to engage an individual academic as a consultant. Many universities have policies in place that allow their academic staff to spend a certain amount of their time, usually 20%, on outside engagements. Faculty consulting in particular appears to play a significant role in allowing firms to exploit academic expertise for developing patentable technologies. Approximately 26% of patents originated by academic inventors are not assigned to universities but to firms (Thursby, Fuller, and Thursby 2009). One reason for this may be that U.S. academics work as consultants for firms during the summer months, and any intellectual property arising from this activity can be legitimately assigned to their commercial partners.

Research services allow firms to exert a certain amount of control over the output to be generated, manage projects to tighter deadlines than research partnerships (discussed below), and retain tighter control of intellectual property generated. From the viewpoint of the firm, academics are expert specialists capable of resolving specific problems arising in development processes or providing scientifically grounded technological advice. Hiring them as consultants can therefore be an attractive alternative to building internal expertise that would involve major investments in human capital. Research services differ from other forms of university-industry relationships in that they mobilize expertise that is commonly held within academic communities (Agrawal and Henderson 2002). Research services can therefore be seen as leveraging "old science" (Rosenberg 1994; Allen 1977; Gibbons and Johnston 1974). They resolve problems and provide improvements rather than suggesting new project ideas or pioneering new design configurations (Utterback 1994; Gibbons 2000).

However, universities do not generally specialize in providing commercial research services. For firms, the special nature of universities as collaborators poses specific challenges but also offers opportunities. Since academics' livelihood does not depend on providing research services, those associated with highly ranked universities in particular tend to be selective with respect to what projects they engage in, and under what conditions. They may be concerned that projects provide little or no scientific novelty and force them to shift their work from their research efforts toward short-term goals (Boyer and Lewis 1984). In addition, these projects are usually highly confidential, limiting the academic partner's ability to extract academic capital by publishing the results.

There are several circumstances in which firms can, however, successfully source research services from academia in spite of the above challenges. First, academics will have an interest in providing research services if they work in scientific or technological areas that are motivated by application or use (Stokes 1997). A considerable amount of research undertaken at universities and other public research organizations is "applied" in the sense that it addresses technical problems and seeks technical solutions (Niiniluoto 1993). Particularly areas that grew out of professional practice, such as the engineering, law, and medicine, have retained a strong connection with issues of application and implementation. This is reflected in the practices deployed, for instance, by engineers compared to scientists (Allen 1977). Second, firms can use consulting engagements as part of a broader relationship with an academic or their research group. For instance, a short-term consulting engagement may provide the opportunity for assessing the potential of more substantial cooperation. Alternatively, within the context of a larger university-firm research alliance, academics may provide contract research or consulting as part of broader terms of mutual exchange (Perkmann and Walsh 2009). Third, academics will be available when firms request services for which the marginal cost to the academic providing it is small. This is the case when firms require relatively rare, highly specialized expertise that is of high value to them but can be offered by an academic relatively easily as he or she possesses the required expertise already and will not have to acquire new knowledge to carry out the assignment. In other words, academics will be well disposed to consult if they can extract a rent from their expertise.

What can firms do in order to successfully source research services from universities? A first aspect is to ensure that transaction costs are held low. Consulting projects, for instance, are relatively small projects with defined outcomes and timelines. As university policies on intellectual property have become increasingly ambitious, negotiations over terms and conditions of the contracts between firms and universities can become arduous and long-lasting. To reduce transaction costs, and minimize the duration of negotiations, actors have sought to create standardized agreements that can be flexibly deployed. For instance, P&G has developed a framework agreement with the State of Ohio, covering all universities in the state system. Similarly, in the U.K., so-called Lambert Agreements serve as a model contracts for university-industry collaboration (Treasury 2003).

A second aspect concerns the need for the projects to be "ambidextrous" in the sense that they fulfill both the firm's and the academic's requirements. Firms experienced in sourcing university knowledge in this way are aware that by allowing their academic partners to simultaneously pursue academic goals—publishing, PhD supervision, access to data or artifacts—they may considerably lower the cost of the engagement or enhance academics' motivation for engaging in the task.

A third aspect is common to R&D outsourcing generally. Firms use research contracts for defining requirements and project outputs whereby compensation is usually agreed on a time-and-effort or cost-plus basis (Carson 2007). As the provision of such services is subject to a relatively high degree of uncertainty, the presence of trust, especially goodwill trust (Sako 1992), will usually be a precondition for successful sourcing of R&D. Research has shown that the formal stipulations of contracts in such circumstances can be "forgotten" if the exigencies of project progress and day-to-day operations demand so (Howells 1999).

Research Partnerships

Research partnerships are a mode of open innovation that is particularly relevant for science-intensive firms across a variety of sectors (Cohen, Nelson, and Walsh 2002; Meyer-Krahmer and Schmoch 1998). In common with research services, research partnerships represent a means for accessing university-generated knowledge even in sectors and areas of expertise where appropriation via intellectual property rights is less effective. In open innovation terms, such partnerships fit the "coupled" path of open innovation where two or more parties contribute to the creation and development of an innovation (Enkel, Gassmann, and Chesbrough 2009).

Research partnerships are partnering arrangements to which both parties bring to bear their assets and competencies. They differ from licensing arrangements in that there is no preexisting body of technology the partners view as commercially valuable enough to justify straight intellectual property transfer. Organizationally, research partnerships range from small-scale ad-hoc collaboration to large, collaborative research centers (Adams, Chiang, and Starkey 2001; Boardman and Gray 2010; Boardman and Corley 2008; Santoro and Chakrabarti 1999; Bozeman and Boardman 2003).

Research partnerships are often subsidized by public policy programs. In the U.S., this is accomplished via federal-funded schemes such as the Advanced Technology Program (ATP) (Hall, Link, and Scott 2000), various funding instruments provided by research councils, government departments, and the National Health Service in the U.K. (Howells, Nedeva, and Georghiou 1998) and joint university-industry projects within federal programs in Germany (Schmoch 1999). In Europe, the "framework programs" of the European Commission provide resources for collaborative projects involving universities and firms (Larédo and Mustar 2004; Caloghirou, Tsakanikas, and Vonortas 2001; Peterson and Sharp 1998), complemented by programs of the national funding councils.

For firms, the availability of public funds amounts to a de-facto subsidy that allows them to leverage its own R&D budget. However, publicly funded programs usually stipulate that for the work undertaken within such partnerships to be "pre-competitive" (i.e., without immediate commercial appropriation potential). The mobilization of public funding means that universities claim full or at least part ownership on intellectual property arising from collaborative projects (Owen-Smith 2005; Ham and Mowery 1998; Poyago-Theotoky, Beath, and Siegel 2002; Fontana, Geuna, and Matt 2006; Caloghirou, Tsakanikas, and Vonortas 2001).

While most research partnerships are established to tackle rather basic science challenges that maybe five to fifteen years removed from commercial application, they differ in terms of intellectual property agreements. One the one hand, there are open "grand challenges" where firms accept that research results are openly publishing, and intellectual property considerations retreat into the background. An example of such an open grand challenge is the Structural Genomics Consortium, an innovative collaboration between pharma companies GlaxoSmithKline, Merck, and Novartis to carry out research on proteins relevant to drug discovery. The consortium, based at Toronto, Oxford, and Stockholm, has attracted funding from the Wellcome Trust and numerous other public funders and foundations. In order to reduce transaction costs and focus the participants' minds on the scientific challenges, the organization has an explicit policy not to pursue any intellectual property rights protection (Perkmann and Schildt 2011).

For firms, participation in such open science initiatives may make sense, for instance when leading industry players become aware that collective action is needed to address fundamental challenges experienced by the entire industry such as in pharmaceuticals where R&D productivity has dramatically dropped. Furthermore, open research partnerships can help firms to create entirely new markets. Knowledge not subject to IP restrictions is likely to diffuse quicker and wider than protected knowledge, and hence draws in a larger number of follow-on researchers and innovators that were not part to the original research program. IBM has sponsored major research initiatives in the emerging field of services science, allowing it to influence research problems and research agendas. A third rationale for participating in open research partnerships is to address challenges of high social relevance, such as in energy or environmental protection. Oil and gas firms, for instance, have funded large university-based programs in alternative energies or carbon capture.

Panagopoulos (2003) proposes an economic model according to which firms are more likely to choose research partnerships when the technology in question is novel and not well understood. Under these conditions, the opportunity cost of disclosing intellectual property and forgoing (some) intellectual property arising from collaborative research are lower and potential gains from knowledge "spillovers" arising from cooperation are higher than in the case of developing and improving mature technologies. Firms will accept such "open" agreements for upstream R&D where outcomes are highly uncertain, time frames are long-term, and the commercial value of the technology in question is not proven.

In contrast to such open partnerships, "closed" partnerships between firms and universities put more emphasis on the protection of property rights. Many firms have in recent years reduced the budgets they allocate to intramural R&D, and have made efforts to in-source more of their R&D activities from other providers (Dodgson, Gann, and Salter 2006). In this exercise, universities represent attractive partners, particularly in areas of high public research investment, such as pharmaceuticals or defense (Garnier 2008). For Rolls-Royce, for instance, collaboration with universities has become a natural complement to their intramural R&D. The firm has established approximately thirty "university technology centers" at different universities, from the U.S. to China. Each center focuses on a specialized area of technology, such as vibration or combustion, and each university partner is chosen for its excellence in an area. The centers are funded for five-year intervals and renewals are linked to specific performance milestones. The UTCs have confidentiality agreements in place, and IP rights are controlled by Rolls-Royce.

Closed partnerships allow a corporate partner closer control of both research program and research outcomes but are likely to attract less public funding particularly if only one firm is involved. A middle way between both extremes is chosen in partnerships with multiple corporate participants where the latter are given the right of first refusal to license any technologies that may arise from the activities of a partnership or a center.

To summarize, research partnerships allow firms to participate in the generation of new knowledge, deploying the talent and the resources available at institutions conducting public research. The fact that particularly in open partnerships costs can be scaled across a number of participants without necessarily reducing the value created is key to understanding the attractiveness of research partnerships. These potential gains provide a counterweight to the uncertainty and appropriability concessions attached to participation in such programs. For instance, a study of SEMATECH, a large research network in the semiconductor industry, showed that the partnership provided for advancements in processes and technologies which would not have been undertaken by any of the parties in isolation (Link, Teece, and Finan 1996).

What capabilities do firms need to engage in research partnerships? Traditional search and monitoring capability, for instance, are of lesser importance when the technology assets to be sourced are yet to be created. Given that learning alliances often have a more emergent quality rather than being accessible to ex-ante matching (Koza and Lewin 1998), rational search routines will be of lesser importance than the actual management of the interorganizational collaboration arrangement. In this respect, both specifying outcomes ex-ante and evaluating outcomes ex-post have been found to be difficult (Hagedoorn, Link, and Vonortas 2000). Therefore, in terms of management, research partnerships rely strongly on the auto-organizing initiative of participating individuals and teams, for instance as "boundary-spanners" acting as brokers between the organization and its environment (Tushman and Scanlan 1981). This will enable an organization to immerse itself in ecologies of knowledge creation by relying on a variety of interpersonal links across its organizational boundary (Liebeskind et al. 1996).

Equally, while licensing constitutes a clearly delineated transaction or series of transactions, the terms of explorative collaboration are more difficult to specify ex ante, particularly in the case of university-industry partnerships (Perkmann, Neely, and Walsh 2011). Control of such collaboration therefore relies on behavioral and process controls rather than output controls (Koza and Lewin 1998). Relying on input and throughput controls, rather than output controls, means such alliances are more unpredictable in their outcomes. Trusted relationships between individual members of the involved organizations are therefore likely to play a more important role than technology trading situations (McEvily, Perrone, and Zaheer 2003; Liebeskind et al. 1996).

Finally, results from partnerships need to be successfully appropriated by the firm (Cohen and Levinthal 1990). As participation in such alliances will most likely be led by specialized R&D personnel, the challenges lies in the how the gained knowledge can be successfully deployed by operational functions within the firm. In other words, the management of interfaces and cross-functional feedback therefore appear crucial. Given the nonlinearity of the innovation process (Kline 1985), firms will require integration mechanisms to successfully exploit newly generated knowledge acquired through explorative alliances.

Under what conditions will this type of open innovation be most effective? Lane and Lubatkin (1998) emphasize that successful learning alliances must fulfill a series of preconditions, such as similarity between the partners' compensation policies and organizational structures, and the partners' familiarity with each others' set of organizational problems. This is why biotechnology firms are often organized similarly to university departments (Zucker and Darby 1997). In the same vein, others have found that pharmaceutical firms were best able to learn from the science community when their organizational structures allowed their scientists to establish network relationships with academic scientists (Cockburn and Henderson 1998; Owen-Smith and Powell 2004). More generally speaking, exploration alliances are most likely to occur in fields that are newly emerging, implying that knowledge is highly dispersed and not yet codified (Powell, Koput, and Smith-Doerr 1996). Similarly, exploration alliances tend to be practiced in innovation areas that are highly dependent on academic science or, alternatively, on the input of dispersed and noncommercially acting users (Von Hippel 1987) or open-source communities (Dahlander and Magnusson 2005).

Conclusions

The three modes of sourcing knowledge outlined above represent different ways of engaging in open innovation by accessing public science. Most evidently, these modes align differently with open science. Licensing represents the standard "outside-in" model of open innovation where technology is packaged into intellectual property and traded for profit outside the channels of open science. Similarly, research services often stipulate confidentiality and firms' control over any prospective intellectual property. By contrast, many research partnerships represent the opposite end of the spectrum as they implicitly or explicitly embrace aspects of open science with its emphasis on open publishing and curiosity-driven research.

Despite different emphases, however, there is no necessary contradiction between any of these modes and open science. Technology transfer via licensing occurs when the science has already been conducted, and in many cases research results have been published in the open literature in parallel to being patented. Research on "paper patent pairs" demonstrates that many scientists are expert in playing in both the academic and commercial fields (Murray 2002) and both types of activities are often complementary (Perkmann, King, and Pavelin 2011). Furthermore, the creation of intellectual property rights often arises from work conducted during research partnerships and research services. Finally, research services are often provided by academics whose inventions have been licensed to firms but require more inventor involvement in order to successfully exploit the technologies in question (Agrawal 2006).

We end by identifying future research opportunities at the intersection of open innovation and university technology transfer. While open innovation scholars examine corporate sourcing of external knowledge, research on sourcing such knowledge specifically from universities is not as abundant (West and Bogers 2014). Similarly, while there is a large body of research on university technology transfer, much of this research is empirically focused on the university side. For these reasons, more evidence and theory development is required on the characteristics, drivers, and consequences of all three models of knowledge sources that we identified above specifically from the vantage point of the firm.

First, there is a need for more research on the costs and benefits of university licensing to firms. In the wake of the Bayh-Dole Act, many universities have—supported by policy makers—put major emphasis on exploiting intellectual property rights. This has led some observers to complain that transaction costs relating to the negotiation of intellectual property rights for research partnerships and research services have become too onerous, deterring firms from engaging with universities, and likewise frustrating academics' attempts to work with industry (Bruneel, D'Este, and Salter 2010; Kenney and Patton 2009). As a number of universities and firms have been experimenting with various approaches to address these issues, future research should assess these partnerships from the firms' viewpoint, and identify their conditions for success.

More generally, we need a better understanding of how sourcing of university knowledge contributes to firm performance. While the literature on R&D alliances and open innovation has provided rich insight into the effects of external knowledge sourcing on performance (Sampson 2007; Laursen and Salter 2006; West and Bogers 2014), we know less about the performance effects of sourcing university knowledge. Specifically, future research should investigate what moderates the benefits of university sourcing strategies.

Second, the research on faculty consulting and contract research has emphasized potential positive and negative impacts on the university, such as possible complementarities with basic research and conflicts of interest between the faculty and consulting roles, respectively. Research on the role of research services for firms is more limited, however. Research services are inherently difficult to study because of confidentiality agreements, and their relatively informal nature, particularly from firms' viewpoint. However, anecdotal insights suggest that there is considerable variation in terms of the possible contradictions and complementarities between private and open science objectives, providing the opportunity to study what moderates the eventual outcomes. The case for further research in this area is strengthened by the fact that firms' expenditure is on average a multiple of their expenditure on securing rights on intellectual property from universities.

Third, more research is needed on how and why firms engage in collaborative research partnerships. For instance, a relevant question pertaining to research partnerships is how they should be structured organizationally so they are not only able to overcome the challenges inherent in university-industry relations but also exploit their unique properties (Cyert and Goodman 1997). An equally important question is how firms should manage their transition to new R&D business models where a larger part of R&D is sourced via partnerships, partly with universities (Garnier 2008). At the same time, more evidence is needed on the new breed of collaborative initiatives that seek to emphasize open science over intellectual property considerations, as seen in the case of the Structural Genomics Consortium.

A final area for future research concerns individual motivations and the conditions under which academic researchers are both prepared and able to participate in corporate research. For firms, understanding the motivations is important to agreeing to partnerships whereby the motivations are often non-pecuniary (D'Este and Perkmann 2011). Firms may be able to use their resources to provide nonmonetary rewards such as support for academic research and the pursuit of recognition within the system of open science, in analogy to the incentives applicable to other types of innovators, such as open-source programmers, students, and/or consumers (West and Gallagher 2006; Dahlander and Gann 2010). The overarching question here is how firms can motivate and access the follow-on innovations that distributed innovators may generate in the wider ecosystem of research and development in which firms operate.

Notes

1. Here we use the term "university" to include all types of public research organizations, including those primarily funded by the government. This may include public research laboratories, research institutes, and other nonprofit or for-profit research organizations.

2. In this chapter, we explicitly focus on incumbent firms and exclude university spin-off companies which are covered separately in Siegel and Wright (2015) in this handbook.

3. The third path of the "inside-out" (or outbound) open innovation applies to *universities* commercializing technology through the creation of spinoff companies. Because our focus is on how *firms* can benefit from university technology, we do not further consider this case here.

References

Adams, J. D., E. P. Chiang, and K. Starkey. 2001. "Industry-university cooperative research centers." *Journal of Technology Transfer* 26 (1–2): 73–86.

Agrawal, A. 2006. "Engaging the inventor: Exploring licensing strategies for university inventions and the role of latent knowledge." *Strategic Management Journal* 27 (1): 63–79.

Agrawal, A., and R. M. Henderson. 2002. "Putting patents in context: Exploring knowledge transfer from MIT." *Management Science* 48 (1): 44–60.

Allen, T. J. 1977. *Managing the flow of technology: Technology transfer and the dissemination of technological information within the R & D organization.* Cambridge, MA: MIT Press.

Argyres, N. S., and J. P. Liebeskind. 1998. "Privatizing the intellectual commons: Universities and the commercialization of biotechnology." *Journal of Economic Behavior and Organization* 35 (4): 427–54.

Arora, A., A. Fosfuri, and A. Gambardella. 2001. "Markets for technology and their implications for corporate strategy." *Industrial and Corporate Change* 10 (2): 419–51.

Arundel, A., and A. Geuna. 2004. "Proximity and the use of public science by innovative European firms." *Economics of Innovation and New Technology* 13 (6): 559–80.

Bera, R. K. 2009. "The story of the Cohen-Boyer patents." *Current Science* 96 (6): 760–63.

Boardman, C., and D. Gray. 2010. "The new science and engineering management: Cooperative research centers as government policies, industry strategies, and organizations." *Journal of Technology Transfer* 35 (5): 445–59.

Boardman, P. C., and E. A. Corley. 2008. "University research centers and the composition of research collaborations." *Research Policy* 37 (5): 900–13.

Boyer, C. M., and D. R. Lewis. 1984. "Faculty consulting: Responsibility or promiscuity?" *Journal of Higher Education* 55 (5): 637–59.

Bozeman, B., and P. C. Boardman. 2003. "Managing the new multipurpose, multidiscipline university research centers: Institutional innovation in the academic community." *Report prepared for the Transforming Organizations Series. IBM Center for the Business of Government.*

Bruneel, J., P. D'Este, and A. Salter. 2010. "Investigating the factors that diminish the barriers to university-industry collaboration." *Research Policy* 39 (7): 858–68.

Brusoni, S., O. Marsili, and A. Salter. 2005. "The role of codified sources of knowledge in innovation: Empirical evidence from Dutch manufacturing." *Journal of Evolutionary Economics* 15 (2): 211–31.

Caloghirou, Y., A. Tsakanikas, and N. S. Vonortas. 2001. "University-industry cooperation in the context of the European framework programmes." *Journal of Technology Transfer* 26 (1–2): 153–61.

Carayol, N. 2003. "Objectives, agreements and matching in science-industry collaborations: Reassembling the pieces of the puzzle." *Research Policy* 32 (6): 887–908.

Carson, S. J. 2007. "When to give up control of outsourced new product development." *Journal of Marketing* 71 (1): 49–66.

Cassiman, B., M. C. Di Guardo, and G. Valentini. 2010. "Organizing links with science: Cooperate or contract? A project-level analysis." *Research Policy* 39 (7): 882–92.

Chesbrough, H. W. 2003. *Open innovation: The new imperative for creating and profiting from technology.* Boston: Harvard Business School Press.

Chok, J. I. 2009. "Regulatory dependence and scientific advisory boards." *Research Policy* 38 (5): 710–25.

Cockburn, I. M., and R. M. Henderson. 1998. "Absorptive capacity, coauthoring behavior, and the organization of research in drug discovery." *Journal of Industrial Economics* 46 (2): 157–82.

Cohen, W. M., and D. A. Levinthal. 1990. "Absorptive capacity: A new perspective on learning and innovation." *Administrative Science Quarterly* 35 (1): 128–52.

Cohen, W. M., R. R. Nelson, and J. P. Walsh. 2002. "Links and impacts: The influence of public research on industrial R&D." *Management Science* 48 (1): 1–23.

Colyvas, J., et al. 2002. "How do university inventions get into practice?" *Management Science* 48 (1): 61–72.

Cyert, R. M., and P. S. Goodman. 1997. "Creating effective university-industry alliances: An organizational learning perspective." *Organizational Dynamics* 25 (4): 45–57.

Dahlander, L., and D. M. Gann. 2010. "How open is innovation?" *Research Policy* 39 (6): 699–709.

Dahlander, L., and M. G. Magnusson. 2005. "Relationships between open source software companies and communities: Observations from Nordic firms." *Research Policy* 34 (4): 481–93.

Dasgupta, P., and P. A. David. 1994. "Toward a new economics of science." *Research Policy* 23 (5): 487–521.

David, P. A. 2004. "Can "Open Science" be protected from the evolving regime of IPR protections?" *Journal of Institutional and Theoretical Economics* 160 (1): 9–34.

Debackere, K. and R. Veugelers. 2005. "The role of academic technology transfer organizations in improving industry science links." *Research Policy* 34 (3): 321–42.

D'Este, P., and P. Patel. 2007. "University-industry linkages in the UK: What are the factors determining the variety of interactions with industry?" *Research Policy* 36 (9): 1295–1313.

D'Este, P., and M. Perkmann. 2011. "Why do academics engage with industry? The entrepreneurial university and individual motivations." *Journal of Technology Transfer* 36 (3): 316–39.

Dodgson, M., D. Gann, and A. Salter. 2006. "The role of technology in the shift towards open innovation: The case of Procter and Gamble." *R&D Management* 36 (3): 333–46.

Enkel, E., O. Gassmann, and H. Chesbrough. 2009. "Open R&D and open innovation: Exploring the phenomenon." *R&D Management* 39 (4): 311–16.

Etzkowitz, H. 2002. *MIT and the Rise of Entrepreneurial Science*. New York: Routledge.

Fabrizio, K. 2006. "The use of university research in firm innovation." In *Open Innovation: Researching a New Paradigm*, edited by H. W. Chesbrough, W. Vanhaverbeke, and J. West, 134–60. Oxford: Oxford University Press.

Faulkner, W., and J. Senker. 1994. "Making sense of diversity: Public-private sector research linkage in three technologies." *Research Policy* 23 (6): 673–95.

Fontana, R., A. Geuna, and M. Matt. 2006. "Factors affecting university-industry R&D collaboration: The importance of screening and signalling." *Research Policy* 35 (1): 309–23.

Garnier, J.-P. 2008. "Rebuilding the R&D engine in big pharma." *Harvard Business Review* 86 (5): 68–77.

George, G., S. A. Zahra, and D. R. Wood. 2002. "The effects of business-university alliances on innovative output and financial performance: A study of publicly traded biotechnology companies." *Journal of Business Venturing* 17 (6): 577–609.

Gibbons, M. 2000. "Changing patterns of university: Industry relations." *Minerva* 38 (3): 352–61.

Gibbons, M., and R. Johnston. 1974. "The roles of science in technological innovation." *Research Policy* 3 (3): 220–42.

Hagedoorn, J., A. N. Link, and N. S. Vonortas. 2000. "Research partnerships." *Research Policy* 29 (4–5): 567–86.

Hall, B. H., A. N. Link, and J. T. Scott. 2000. "Universities as research partners." *Review of Economics and Statistics*, 85: 485–91.

Hall, B. H., A. N. Link, and J. T. Scott. 2001. "Barriers inhibiting industry from partnering with universities: Evidence from the Advanced Technology Program." *Journal of Technology Transfer* 26 (1): 87–98.

Ham, R. M., and D. C. Mowery. 1998. "Improving the effectiveness of public-private R&D collaboration: Case studies at a US weapons laboratory." *Research Policy* 26 (6): 661–75.

Hanel, P., and M. St-Pierre. 2006. "Industry-university collaboration by Canadian manufacturing firms." *Journal of Technology Transfer* 31 (4): 485–99.

Henderson, R., A. B. Jaffe, and M. Trajtenberg. 1998. "Universities as a source of commercial technology: A detailed analysis of university patenting 1965–1988." *Review of Economics and Statistics* 80 (1): 119–27.

Howells, J. 1999. "Research and technology outsourcing." *Technology Analysis and Strategic Management* 11 (1): 17–29.

Howells, J., M. Nedeva, and L. Georghiou. 1998. *Industry-academic links in the UK*. Manchester PREST, University of Manchester.

Hülsbeck, M., and E. E. Lehmann. 2012. "Academic entrepreneurship and board formation in science-based firms." *Economics of Innovation and New Technology* 21 (5–6): 547–65.

Jensen, R., and M. Thursby. 2001. "Proofs and prototypes for sale: The licensing of university inventions." *American Economic Review* 91 (1): 240–59.

Kenney, M. 1986. *Biotechnology: The University-Industrial Complex*. New Haven, CT: Yale University Press.

Kenney, M., and D. Patton. 2009. "Reconsidering the Bayh-Dole Act and the current university invention ownership model." *Research Policy* 38 (9): 1407–22.

Kline, S. J. 1985. "Innovation is not a linear process." *Research Management* 28 (4): 36–45.

Koza, M. P., and A. Y. Lewin. 1998. "The co-evolution of strategic alliances." *Organization Science* 9 (3): 255–64.

Krimsky, S. 2003. *Science in the Private Interest: Has the Lure of Profits Corrupted the Virtue of Biomedical Research?* Lanham, MD: Rowman and Littlefield.

Lane, P. J., and M. Lubatkin. 1998. "Relative absorptive capacity and interorganizational learning." *Strategic Management Journal* 19 (5): 461–77.

Larédo, P., and P. Mustar. 2004. "Public sector research: A growing role in innovation systems." *Minerva* 42 (1): 11–27.

Laursen, K., and A. Salter. 2004. "Searching high and low: What types of firms use universities as a source of innovation?." *Research Policy* 33 (8): 1201–15.

Laursen, K., and A. Salter. 2006. "Open for innovation: the role of openness in explaining innovation performance among U.K. manufacturing firms." *Strategic Management Journal* 27 (2): 131–50.

Lee, Y. S. 2000. "The sustainability of university-industry research collaboration: An empirical assessment." *Journal of Technology Transfer* 25 (2): 111–33.

Lenoir, T. 1997. *Instituting science: The cultural production of scientific disciplines*. Stanford, CA: Stanford University Press.

Levin, R. C., et al. 1987. "Appropriating the returns from industrial research and development." *Brookings Papers on Economic Activity* 1987 (3): 783–820.

Liebeskind, J. P., et al. 1996. "Social networks, learning, and flexibility: Sourcing scientific knowledge in new biotechnology firms." *Organization Science* 7 (4): 428–43.

Link, A. N., D. S. Siegel, and B. Bozeman. 2007. "An empirical analysis of the propensity of academics to engage in informal university technology transfer." *Industrial and Corporate Change* 16 (4): 641–55.

Link, A. N., D. J. Teece, and W. F. Finan. 1996. "Estimating the benefits from collaboration: The case of SEMATECH." *Review of Industrial Organization* 11 (5): 737–51.

Lockett, A., and M. Wright. 2005. "Resources, capabilities, risk capital and the creation of university spin-out companies." *Research Policy* 34 (7): 1043–1057.

MacPherson, A. 2002. "The contribution of academic-industry interaction to product innovation: The case of New York State's medical devices sector." *Papers in Regional Science*, vol. 81: 121–29.

Mansfield, E. 1991. "Academic research and industrial innovation." *Research Policy* 20 (1): 1–12.

McEvily, B., V. Perrone, and A. Zaheer. 2003. "Trust as an organizing principle." *Organization Science* 14 (1): 91–103.

Merton, R. K. 1973. *The sociology of science: Theoretical and empirical investigations*. Chicago: University of Chicago Press.

Meyer-Krahmer, F., and U. Schmoch. 1998. "Science-based technologies: University-industry interactions in four fields." *Research Policy* 27 (8): 835–51.

Mohnen, P., and C. Hoareau. 2003. "What type of enterprise forges close links with universities and government labs? Evidence from CIS 2." *Managerial and Decision Economics* 24 (2–3): 133–45.

Mowery, D. C. 2009. "Plus ca change: Industrial R&D in the "third industrial revolution." *Industrial and Corporate Change* 18 (1): 1–50.

Mowery, D. C., and B. N. Sampat. 2005. "Universities in national innovation systems." In *The Oxford Handbook of Innovation*, edited by J. Fagerberg, D. Mowery, and R. Nelson, 209–39. Oxford: Oxford University Press.

Murmann, J. P. 2003. *Knowledge and competitive advantage: The coevolution of firms, technology, and national institutions*. Cambridge: Cambridge University Press.

Murray, F. 2002. "Innovation as co-evolution of scientific and technological networks: Exploring tissue engineering." *Research Policy* 31 (8–9): 1389–1403.

Murray, F. 2010. "The oncomouse that roared: Hybrid exchange strategies as a source of distinction at the boundary of overlapping institutions." *American Journal of Sociology* 116 (2): 341–88.

Murray, F., and S. Stern. 2007. "Do formal intellectual property rights hinder the free flow of scientific knowledge? An empirical test of the anti-commons hypothesis." *Journal of Economic Behavior and Organization* 63 (4): 648–87.

Nelson, R. R. 2001. "Observations on the post-Bayh-Dole rise of patenting at American universities." *Journal of Technology Transfer* 26 (1–2): 13–19.

Niiniluoto, I. 1993. "The aim and structure of applied research." *Erkenntnis* 38 (1): 1–21.

Niosi, J. 2006. "Introduction to the symposium: Universities as a source of commercial technology." *Journal of Technology Transfer* 31 (4): 399–402.

Owen-Smith, J. 2005. "Trends and transitions in the institutional environment for public and private science." *Higher Education* 49 (1): 91–117.

Owen-Smith, J., and W. W. Powell. 2001. "To patent or not: Faculty decisions and institutional success at technology transfer." *Journal of Technology Transfer* 26 (1): 99–114.

Owen-Smith, J., and W. W. Powell. 2004. "Knowledge networks as channels and conduits: The effects of spillovers in the Boston biotechnology community." *Organization Science* 15 (1): 5–21.

Panagopoulos, A. 2003. "Understanding when universities and firms form RJVs: The importance of intellectual property protection." *International Journal of Industrial Organization* 21 (9): 1411–33.

Pavitt, K. 1991. "What makes basic research economically useful?" *Research Policy* 20 (2): 109–19.

Perkmann, M., Z. King, and S. Pavelin. 2011. "Engaging excellence? Effects of faculty quality on university engagement with industry." *Research Policy* 40 (4): 539–52.

Perkmann, M., A. Neely, and K. Walsh. 2011. "How should firms evaluate success in university-industry alliances? A performance measurement system." *R & D Management* 41 (2): 202–16.

Perkmann, M., and H. Schildt. 2011. "Open data in industrial R&D: Organizing radically open collaboration between firms and public science," *Discussion Paper, Imperial College Business School.*

Perkmann, M., and K. Walsh. 2007. "University-industry relationships and open innovation: Towards a research agenda." *International Journal of Management Reviews* 9 (4): 259–80.

Perkmann, M., and K. Walsh. 2008. "Engaging the scholar: Three forms of academic consulting and their impact on universities and industry." *Research Policy* 37 (10): 1884–91.

Perkmann, M., and K. Walsh. 2009. "The two faces of collaboration: Impacts of university-industry relations on public research." *Industrial and Corporate Change* 18 (6): 1033–65.

Peterson, J., and M. Sharp. 1998. *Technology policy in the European Union.* Basingstoke, U.K.: Macmillan.

Phan, P. H., and D. S. Siegel. 2006. "The effectiveness of university technology transfer: Lessons learned from quantitative and qualitative research in the U.S. and the U.K." *Rensselaer Working Papers in Economics.* Troy, NY.

Powell, W. W., K. W. Koput, and L. Smith-Doerr. 1996. "Interorganizational collaboration and the locus of innovation: Networks of learning in biotechnology." *Administrative Science Quarterly* 41 (1): 116–45.

Poyago-Theotoky, J., J. Beath, and D. S. Siegel. 2002. "Universities and fundamental research: Reflections on the growth of university-industry partnerships." *Oxford Review of Economic Policy* 18 (1): 10–21.

Rivette, K. G., and D. Kline. 2000. "Discovering new value in intellectual property." *Harvard Business Review* 78 (1): 54–66.

Roessner, J. D. 1993. "What companies want from the federal labs." *Issues in Science and Technology* 10 (1): 37–42.

Rosenberg, N. 1994. *Exploring the black box: Technology, Economics, and History.* Cambridge: Cambridge University Press.

Rosenkopf, L., and A. Nerkar. 2001. "Beyond local search: Boundary-spanning, exploration, and impact in the optical disk industry." *Strategic Management Journal* 22 (4): 287–306.

Sako, M. 1992. *Prices, quality, and trust: Inter-firm relations in Britain and Japan*. Cambridge: Cambridge University Press.

Salter, A. J., and B. R. Martin. 2001. "The economic benefits of publicly funded basic research: A critical review." *Research Policy* 30 (3): 509–32.

Sampson, R. C. 2007. "R&D alliances and firm performance: The impact of technological diversity and alliance organization on innovation." *Academy of Management Journal ARCHIVE* 50 (2): 364–86.

Santoro, M. D., and A. K. Chakrabarti. 1999. "Building industry-university research centers: Some strategic considerations." *International Journal of Management Reviews* (3): 225–44.

Sapsalis, E., B. Van Pottelsberghe De La Potterie, and R. Navon. 2006. "Academic versus industry patenting: An in-depth analysis of what determines patent value." *Research Policy* 35 (10): 1631–45.

Schartinger, D., et al. 2002. "Knowledge interactions between universities and industry in Austria: Sectoral patterns and determinants." *Research Policy* 31 (3): 303–28.

Schmoch, U. 1999. "Interaction of universities and industrial enterprises in Germany and the United States: A comparison." *Industry and Innovation* 6 (1): 51–68.

Shane, S. A. 2005. *Economic Development through Entrepreneurship: Government, University and Business Linkages*. Cheltenham, U.K.: Edward Elgar.

Siegel, D. S., D. Waldman, and A. Link. 2003. "Assessing the impact of organizational practices on the relative productivity of university technology transfer offices: An exploratory study." *Research Policy* 32 (1): 27–48.

Siegel, D., and M. Wright. 2015. "University Technology Transfer Offices, Licensing, and Start-Ups." In *Chicago Handbook of University Technology Transfer*. Chicago: University of Chicago Press.

Slaughter, S., and L. L. Leslie. 1997. *Academic Capitalism: Politics, Policies and the Entrepreneurial University*. Baltimore, MD: Johns Hopkins University Press.

Stokes, D. E. 1997. *Pasteur's Quadrant: Basic Science and Technological Innovation*. Washington, D.C.: Brookings Institution Press.

Stuart, T. E., and W. W. Ding. 2006. "When do scientists become entrepreneurs? The social structural antecedents of commercial activity in the academic life sciences." *American Journal of Sociology* 112 (1): 97–144.

Stuart, T. E., S. Z. Ozdemir, and W. W. Ding. 2007. "Vertical alliance networks: The case of university-biotechnology-pharmaceutical alliance chains." *Research Policy* 36 (4): 477–98.

Teece, D. J. 1986. "Profiting from technological innovation: Implications for integration, collaboration, licensing and public policy." *Research Policy* 15 (6): 285–305.

Thursby, J., A. Fuller, and M. Thursby. 2009. "US faculty patenting: Inside and outside the university." *Research Policy* 38 (1): 14–25.

Thursby, J. G., and M. C. Thursby. 2001. "Industry perspectives on licensing university technologies: Sources and problems." *Industry and Higher Education* 15 (4): 289–94.

Thursby, J. G., and M. C. Thursby. 2004. "Are faculty critical? Their role in university-industry licensing." *Contemporary Economic Policy*, 22 (2): 162–78.

Thursby, J. G. A., R. A. Jensen, and M. C. A. Thursby. 2001. "Objectives, characteristics and outcomes of university licensing: A survey of major US universities." *Journal of Technology Transfer* 26 (1): 59–72.

Treasury, H. 2003. *Lambert Report of University-Industry Collaboration: Final Report*. London: Her Majesty's Stationery Office.

Tushman, M. L., and T. J. Scanlan. 1981. "Boundary spanning individuals: Their role in information transfer and their antecedents." *Academy of Management Journal* 24 (2): 289–305.

Utterback, J. M. 1994. *Mastering the Dynamics of Innovation: How Companies Can Seize Opportunities in the Face of Technological Change*. Boston: Harvard Business School Press.

Von Hippel, E. 1987. *The Sources of Innovation*. New York: Oxford University Press.

West, J., and M. Bogers. 2014. "Leveraging external sources of innovation: A review of research on open innovation." *Journal of Product Innovation Management* 31 (4): 814–31.

West, J., and S. Gallagher. 2006. "Challenges of open innovation: the paradox of firm investment in open-source software." *R&D Management* 36 (3): 319–31.

White, J. T., et al. 2013. Appointments of Academic Directors, DERA Working Paper 2013–02, Security and Exchange Commission. https://http://www.sec.gov.edgekey.net/divisions/riskfin/workingpapers/rsfi-wp2013–02.pdf.

Winter, S. G. 1987. "Knowledge and competence as strategic assets: Strategies for industrial innovation and renewal." In *The Competitive Challenge*, edited by D. J. Teece, 159–84. Cambridge, MA: Ballinger.

Zucker, L. G., and M. R. Darby. 1997. "Individual action and the demand for institutions: Star scientists and institutional transformation." *American Behavioral Scientist* 40 (4): 502–13.

Zucker, L. G., M. R. Darby, and J. S. Armstrong. 2002. "Commercializing knowledge: University science, knowledge capture, and firm performance in biotechnology." *Management Science* 48 (1): 138–53.

TABLE 2.2. **Sourcing knowledge from universities: Summary of previous research.**

Article	Research Questions	Data	Findings
Adams, Chiang, and Starkey, 2001.	What is the influence of university-industry centers on industrial R&D and patenting activity?	Survey of 208 R&D labs in U.S. owned by 200 public companies	Analyzing university-industry research centers across all industries, the authors provide evidence that they promote technology transfer by increasing patenting rates at the associated industrial laboratories. In addition to classic intellectual property transfer, the centers stimulate a range of activities such as coauthoring between university and industry members, academic consulting, applied R&D, educational outputs.
Agrawal 2006.	To what extent is the commercialization of university technology dependent on the involvement of the inventor?	438 license agreements from Massachusetts Institute of Technology	Inventor involvement is positively related to commercialization success although a number of caveats are highlighted.
Arundel and Geuna 2004.	To what degree do large firms use knowledge from public science organizations, and what role does proximity play?	Survey data of large R&D-based EU firms; 588 responses.	Find that knowledge from public research organizations is less important to managers, than knowledge from affiliated firms. They find *domestic*-based public science is positively correlated with the quality of home science base, and availability of outputs, and negatively correlated with firms' R&D intensity.
Bruneel, D'Este, and Salter 2010.	What reduces the barriers to university-industry collaboration?	Survey of organizations collaborating with U.K. academic researchers in the physical and engineering sciences; responses from 600 organizations.	They identify two types of barriers: barriers due to the basic research orientation of the universities, and transactive barriers. Firms' prior experience of collaborative research lowers orientation-related barriers and greater levels of trust reduce both types of barriers. The breadth of interaction diminishes the orientation-related, but increases transaction-related barriers.

(*continued*)

TABLE 2.2. (*continued*)

Article	Research Questions	Data	Findings
Caloghirou, Tsakanikas, and Vonortas 2001.	Why do firms participate in EU-funded collaborative projects with universities?	Survey of firms participating in EU programs; 312 responses.	Firms collaborate with universities to exploit research synergies (from cost savings or improvements in R&D productivity) and stay current on major technological developments. The most important benefit from such collaboration reported by firms is positive impact on their knowledge base.
Carayol 2003.	What are the matching processes that bring together firms and universities?	Data on 46 collaborations within an EU program	Five types of collaboration are identified, that are differentiated by the type of research, number of participants, age of firms (start-up or not), and organizational structure (formal vs. informal). The author further finds that excellence-seeking academics, focusing on basic research, tend to work with firms pursuing highly risky research.
Cassiman, Di Guardo, and Valentini 2007.	How are university-industry projects structured?	Data on 52 projects involving a multinational firm and partner universities	The authors argue that knowledge characteristics explain how projects are organized. While basic and less strategically important projects are pursued via *cooperative agreements* with universities, for strategically more important projects and for those where the knowledge to be developed is novel to the firm, the firm will use *formal contracting* with a university.
Cockburn and Henderson 1998.	To what degree do firm connections to open science researchers impact the productivity of corporate researchers?	Case histories of 21 high-impact therapeutics; from 81,574 papers published 1980–1994, the coauthorship patterns of researchers at 20 R&D intensive pharma companies	Firms supporting publication in open science are much more likely to collaborate with universities, and such collaboration appears to increase the ability to produce "important" patents.

Cohen, Nelson, and Walsh 2002.	How do firms work with public research organizations, and what is the impact of these links?	Survey of R&D managers in R&D units of U.S. manufacturing firms; 1478 responses.	Although used in a number of manufacturing sectors, public research is critical for a small number of industries (e.g., pharma). In addition to generating new ideas for industrial R&D projects, public research also contributes to the completion of existing ones. Among the most important knowledge pathways are publications and reports, informal information exchange, public meetings and consulting; others are considered less relevant. The most important channels are hence those aligned with open science. The impact of public science industrial R&D is disproportionately larger for larger firms and start-ups.
Colyvas et al. 2002.	What is the impact of intellectual property rights on the commercialization of university inventions?	Eleven case studies of the commercialization of university inventions.	IPR is most important for embryonic inventions, and unimportant for inventions that are useful to industry "off the shelf." Exclusive licenses for embryonic technologies face a high probability that the licensee fails to commercialize it (selection problem). TTOs are most important in areas where existing links between academia and industry are weak.
Cyert and Goodman 1997.	What makes university-industry alliances successful?	Practitioner-oriented article, not based on systematic data set.	The authors argue that research and practice overemphasize effectiveness measures and instead propose a "learning" perspective. Even though university-industry alliances may result in directly commercially relevant outputs, they argue that their more important role is in stimulating interorganizational learning. They conclude that alliances should select problems that are of interest to all parties, pursue team-based collaboration, and create multiple relationships and tasks throughout the life of the alliance.
D'Este and Perkmann 2011.	What motivates academics to engage with industry?	Survey of 4,337 university researchers in the U.K.; 1,528 answered questionnaires.	Most academics engage with industry to further their research rather than to commercialize their knowledge. Patenting and spin-off company formation are motivated exclusively by commercialization while joint research, contract research, and consulting are strongly informed by research-related motives.

(*continued*)

TABLE 2.2. *(continued)*

Article	Research Questions	Data	Findings
Fabrizio 2006.	To what degree do university attempts to commercialize IP slow the availability of public science?	Citation to public science in patents filed by 6,090 U.S. firms from 1976 to 1995	As public science became more important in a technology class, firms separated into "haves" and "have nots" in their access to public science, suggesting firms have unequal access to university research. Also, as university patenting increased, it increased the delays by firms in citing open science, suggesting a slowing down of firm exploitation of existing knowledge.
Faulkner and Senker 1994.	Why do firms work with universities, and what channels do they use?	Qualitative evidence on firms biotechnology, ceramics and parallel computing.	The authors distinguish between three channels used by firms to access university-generated resources: literature, personal contacts and recruitment. They note that most companies engage with a university via informal consulting while formalized relationships are confined to biotechnology and larger firms. In some sectors, such as information technology, university laboratories act more as customers than research partners.
Fontana, Geuna, and Matt 2006.	What determines firms' engagement in research cooperation with public research organizations?	Survey of EU firms in five sectors; 558 responses.	Firms' propensity to collaborate with public research organizations is positively related to the extent to which they screen academic publications, outsource R&D, and patent to protect innovations and signal competencies is positively.
George, Zahra, and Wood 2002.	How does firm collaboration with university impact its innovation and performance measures?	Secondary data on the 2,457 technology alliances of 147 publicly traded publicly traded U.S. biotechnology companies	Firms with university linkages reported more technology alliances, obtain more patents and have lower R&D expenses. Controlling for firm and university competence, more university linkages increase performance, as do linkages with Carnegie Research I universities. Linkages do not increase the number of products developed or released.
Hall, Link, and Scott 2000.	What determines the success of public-private R&D partnerships?	Survey data on 47 projects funded by the U.S. Advanced Technology Program	Projects with university involvement are less likely to terminate early, compared to projects without universities, because university involvement leads to greater awareness of research problems. Firms include universities in projects that involve "new science" (i.e., projects perceived as being problematic with regards to use of basic knowledge).

Hall, Link, and Scott 2001.	What barriers prevent firms from partnering with universities in projects funded by the U.S. Advanced Technology Program?	Survey of participants in 38 Advanced Technology Projects	Projects encountering problems are smaller, shorter, and have a higher ATP share of total funding. Barriers are enhanced by previous experience of working with a university. There are no notable differences between technology areas. Specific difficulties in negotiating intellectual property among the partners are predicted by the ATP share in the project, the lead participant's prior experience with university partnering and being a project in the chemicals industry.
Ham and Mowery 1998.	What makes collaboration between firms and public research organizations effective?	Qualitative evidence on five collaborations (CRADAs) between firms and a U.S. public R&D laboratory. CRADAS are "Cooperative Research and Development Agreements."	Firms' key motive for engaging is access to a unique bundle of interdisciplinary expertise and facilities. Among the problems reported were delays in projects approvals, the difficulties encountered by government laboratory researchers to gather support for these relatively small-scale projects within their organizations as well as limited in-house resources within firms to "absorb" the research results. Also, different "research styles" led to tensions across project teams. The authors conclude that the benefits derived from these projects were mostly generic and could not be specified in terms of direct benefits.
Hanel and St-Pierre 2006.	What are the factors determining firms' collaboration with universities?	Data from Canadian firms included in 1999 Statistics Survey of Innovation, 4,244 observations	The probability of collaborating with a university is positively associated with firm size, being in knowledge-based industries and the importance attributed by a firm to R&D. The latter indicates that university collaboration complements R&D. Collaboration is also positively associated with performance.
Henderson, Jaffe, and Trajtenberg 1998.	Has increased university patent propensity decreased the quality of university patents?	A comparison of 12,804 university patents and 19,535 U.S. PTO patents from 1965–1992, and their respective citations.	The importance and basic-ness of university patents fell dramatically after 1982. The decline is attributed to an increasing fraction of patents from smaller universities, and that all universities are producing more patents that are never cited by anyone.

(*continued*)

TABLE 2.2. (*continued*)

Article	Research Questions	Data	Findings
Jensen and Thursby 2001.	What inventions are universities licensing, and how are they licensing them?	A survey of 62 U.S. research university TTOs	More than 70% of inventions were so early stage that inventor cooperation was required, and 84% of licenses include ongoing revenues. On average, each university's five most important inventions account for 78% of gross license revenue.
Laursen and Salter 2004.	Which firms will draw on university knowledge in their innovation activities?	Responses from 2,655 U.K. manufacturing firms answering the U.K. Innovation Survey	The firms most likely to use university knowledge are those with high R&D intensity (i.e., high absorptive capacity) as well as larger firms overall.
Lee 2000.	Why does industry collaborate with universities occur? What are the resulting benefits?	Survey of U.S. firms affiliated with the Association of University Technology Managers (AUTM); 140 responses.	Firms collaborate with universities to conduct research around existing product lines, pursue exploratory research in search of new products, design of instrumentation, technical problem solving and design of prototypes. Basic research without specific applications is seen as less important. Most firms are satisfied with the outcomes of collaboration, even though faculty contribution to firms' R&D agenda is seen as inconsequential by most firms.
Liebeskind et a1. 1996.	How do biotechnology firms source scientific knowledge?	Longitudinal data on two biotechnology firms, covering formal inter-organizational relationships and bibliographic data	The authors argue that most knowledge transfer between biotechnology firms and academic researchers happens within noncontractual relationships. Knowledge between firms and universities is exchanged within social networks based on trustworthy relationships, and to a lesser degree through formal inter-organizational alliances.
MacPherson 2002.	How does geographic proximity impact academic-industry linkages?	Survey of 63 specialized producers in NY State.	Innovation rates are higher among firms that exploit university resources. Radical innovators have more academic linkages than incremental innovators and attach more importance to proximity.
Mansfield 1991.	To what extent are technological innovations based on recent academic research?	Survey responses from 76 major U.S. firms in seven industrial sectors	Approximately one tenth of new products and processes could not have been developed without substantial detail in the absence of academic research. The average time lag between conclusion of academic research and the launch of innovations is seven years. The estimated social rate of return from academic research is 28%.

Mohnen and Hoareau 2003.	What are the factors determining firms' collaboration with universities and government laboratories?	Community Innovation Surveys (CIS2) for France, Germany, Ireland, Spain.	R&D-intensive firms and radical innovators source knowledge from universities and government labs but not to cooperate with them directly. Outright collaboration with universities and government laboratories is positively associated with firm size, patenting rates, and receipt of government support for innovation.
Owen-Smith, and Powell 2004.	How does knowledge flow through inter-organizational networks?	Data on inter-organizational collaborations involving 482 biotechnology firms in the Boston area	Geographic propinquity and institutional characteristics of key members of a network transform the way in which an organization's position within the network transforms into innovation benefits. Closeness and involvement with public research organizations has positive effects on innovativeness, indicating the impact of informal ties and knowledge spillovers. This effect is context-dependent. In stable environments, closed conduits (formal inter-organizational partnerships) are sufficient while in more variable environments (new sectors) loose "channels" are of benefit for innovators.
Panagopoulos 2003.	Under what conditions do firms engage in Research Joint Ventures with universities?	Presents a economic model	Firms which work on new technologies, are more likely to form partnerships with universities. These firms optimally choose minimal IP protection which lowers profits but also allows them to benefit from increased knowledge spillovers. By contrast, the opportunity cost of joining an RJV for firms developing mature technologies is greater, making such partners unlikely candidates for RJVs.
Roessner 1993.	What benefits do firms expect from working with public research organizations?	Survey of 68 U.S. firms	Analyzes different types of interactions between firms and U.S. federal research labs. The category judged most important by firms was contract research, followed by cooperative research while licensing and formal interactions were judged less important. The major expected benefit for firms was access to technological resources while there was less focus on technology commercialization.

(*continued*)

TABLE 2.2. (*continued*)

Article	Research Questions	Data	Findings
Santoro and Chakrabarti 2002.	What is the impact of firm size on the structure and objectives of partnerships with universities?	Multi-method field study of 21 research centers at U.S. universities.	Larger firms in resource intensive industries use university linkages to build competencies in non-core technology areas. Smaller firms, particularly those in high-tech sectors, focus more on problem solving in core areas, using technology transfer and cooperative research relationships. Individual technology champions at firms play a key role in partnerships with universities.
Sapsalis, Van Pottelsberghe, and Navon 2006.	Are academic patents more or less valuable than industry ones?	An analysis of 400 biotech patents filed by Belgian universities	For both university and corporate patents, patents are more valuable if the author has also contributed to open science, or if the patent cites patents from public institutions.
Thursby and Thursby 2001.	How do firms identify university technologies to license?	A survey of 300 U.S. business units that license external technology	The most important source of licensing opportunities is ongoing contact between corporate researchers and their university counterparts; next most important are academic publications, presentations, and patent filings.
Thursby and Thursby 2004.	What is the role of academic faculty in university-industry licensing deals?	Survey of 112 U.S. firms.	Approx. 40% of all licenses require faculty involvement. For firms, sponsored research can be a substitute for a licensing deal when technologies are too embryonic to be licensed or when the technology involves a platform development or process improvement. Regarding the factors determining success, faculty research orientation (basis/applied) is less important than university receptivity to industrial funding. The authors identify three ways of faculty involvement: identifying technology, developing embryonic technology (not yet licensed), and developing inventions when licensed.
Zucker, Darby, and Armstrong 2002.	How does collaboration with university scientists affect firms' innovativeness?	ISI bibliographic data on publications by academics scientist and scientists associated with selected biotechnology firms	The innovativeness of biotechnology firms is positively correlated with collaboration (indicated by coauthorship) of the firms' scientists with academic star scientists. The authors conclude that the success of firms is not the result of generic knowledge spillovers but linked to actual collaboration with academics that allows the exchange of tacit knowledge.

CHAPTER THREE

Accountability, Government Rights, and the Public Interest

A Thirty-Year Retrospective

Arti Rai and Bhaven Sampat

* We thank Jerry Thursby and participants in the National Academy of Sciences "30 Years After Bayh-Dole" workshop for comments on a previous draft. Ryan O'Quinn and Siobhan Innes-Gawn provided superb research assistance.

1. Introduction

The sponsors of the Bayh-Dole Act believed that giving academic grantees ownership of patent rights on federally funded inventions would be the most effective mechanism for transforming academic discoveries into commercial products (U.S. Senate 1979). Bayh-Dole does not, however, confer entirely unfettered discretion upon grantees. It includes provisions for diligent reporting and retained government rights that can be viewed as "public interest" provisions. Indeed, in *Stanford v. Roche*, a 2011 Supreme Court case, the U.S. government invoked as an argument in favor of default university ownership under Bayh-Dole the claim that such ownership would be more compatible with these public interest provisions than ownership by individual scientists.

In this paper, we examine the role that these provisions have actually played in the 30-year implementation of the statute. Our study proceeds in three parts: first, after a brief discussion of the legislative history of the public interest provisions, we discuss the Department of Commerce's implementing regulations and available case law on sanctions for noncompliance with reporting requirements. We conclude this part by noting the sparse empirical literature on compliance. Next, we turn to large-scale quantitative data that gives some insight into compliance (or lack thereof). Because the largest amount of data on this question is found in the biomedical arena, we focus on this arena. In the third part, we perform a qualitative review focused on drug development, an area where accountability is arguably especially important. Because the universe of academic drug patents is relatively small, we can provide specifics regarding why we believe federal funding was not appropriately reported for at least 15 patents (associated with eight drugs). In addition, for five patents (associated with five drugs), federal funding was involved in at least a portion of the relevant research or in research on compounds very closely related to the patented drug.

2. The Public Interest Provisions of Bayh-Dole: History and Administrative Implementation

The Statute's Public Interest Provisions

Although Bayh-Dole allows academic grantees to own federally funded inventions, these ownership claims are subject to a variety of retained government rights. For example, under Section 203(1) of the statute (35 U.S.C § 203(1)), a federal funding agency may require additional licensing by a grantee or its licensee. The agency can impose additional licensing requirements (sometimes called "march-in" requirements) in situations where the grantee or licensee is not taking effective steps to "achieve practical application" of the invention or where additional licensing is necessary "to alleviate health and safety needs which are not reasonably satisfied." Section 202(c)(4) of the act provides that the agency also has a nonexclusive, paid-up license to practice an invention or have it practiced for, or on behalf of, the U.S. government.

Although federal funding agencies are responsible for exercising these retained rights, the sponsors of Bayh-Dole explicitly contemplated third parties playing a role in assisting these agencies. Specifically, according to the Senate report accompanying S. 414 (the bill that became Bayh-Dole), "complaints from third-parties [would] be the basis for the initiation of agency action."

Bayh-Dole also has a number of reporting requirements that are intended, at least in part, to ensure that agencies and third parties have the information necessary to be aware of retained rights. Section 202(c) requires that the grantee report to the U.S. funding agency any patent application the grantee files and also requires the application to state that the government retains certain rights as a consequence of this federal funding. This section also allows agencies to require "periodic reporting" on utilization or efforts at achieving utilization.

Department of Commerce Regulations

Section 206 of Bayh-Dole gives the Department of Commerce authority to promulgate regulations under Sections 202–4 of the statute. These Commerce regulations are noteworthy because they tend to curtail the effective power of the legislation's public interest provisions. In some circumstances, they do so to an extent that could be challenged as contrary to the statute.

For example, the Commerce Department's regulations implementing march-in rights are quite cumbersome. The march-in provision of the Bayh-Dole Act itself simply states that some sort of procedure for appealing adverse determinations should be available. By contrast, under the implementing regulations, the funding agency must first notify the contractor (and await a reply) before it even begins "considering the exercise of march-in rights." Once the agency has announced that it is "considering" march-in, it must then await the receipt of comments from the contractor and/or licensee. If the comments raise a factual question, the agency must then conduct a formal fact-finding proceeding that affords the contractor "the opportunity to appear with counsel, submit documentary evidence, present witnesses and confront such persons as the agency may present" (37 C.F.R Section 401.6 [e]). Finally, the regulations provide that any agency determination unfavorable to the contractor or licensee be held in abeyance pending the exhaustion of judicial appeals.

This last provision appears directly contrary to the language of the Bayh-Dole statute, which (reasonably enough) does *not* require that march-in be stayed when a health or safety need is "not reasonably satisfied." The requirement for a formal fact-finding proceeding also appears contrary to certain language in the legislative history of S. 414, which notes that "adherence to Administrative Procedures [sic] Act procedures is not required because of concerns that this could frustrate the effectuation of the march-in remedy."

Sanctions for Noncompliance with Reporting Requirements

The issue of sanctions for failure to comply with the reporting requirements of Bayh-Dole has, on occasion, been raised in patent infringement actions. In the 2007 case, *Central Admixture PharmacyServs., Inc. v. Advanced Cardiac Solutions*, P.C., the Court of Appeals for the Federal Circuit (which hears all appeals in patent cases) addressed the defendants' contention that the patentee did not actually have rights in his patent because he had failed to comply fully with relevant federal reporting requirements. The court determined that the funding agency rather than a third party has the right to assert title and invoke forfeiture. Government agencies have, on occasion, successfully taken action to assert title and invoke forfeiture. For example, in the 2004 case of *Campbell Plastics Eng'g and Mfg., Inc. v. Brownlee*, the U.S. Army successfully took title when a government contractor failed to disclose a gas mask invented under a contract to the army.

A few district courts have analyzed the related, but distinct, question of whether failure to disclose government funding in a patent application might constitute "inequitable conduct" before the U.S. Patent and Trademark Office (henceforth, PTO). The sanction for inequitable conduct is unenforceability of the patent. In one case, *McKesson Info. Solutions, LLC v. Trizetto Group, Inc.*, the court ruled that the inequitable conduct allegations were sufficiently pleaded so that the issue could go to trial. The case later settled.

Existing Empirical Research on Compliance

Aside from these legal cases, the literature on compliance is relatively sparse. The relevant data that exists resides in Interagency Edison (iEdison), which incorporates information on patenting and licensing provided by grantees to 29 funding agencies (GAO 2003). However, external researchers are denied access to the iEdison database. This denial of access appears to be based on a Department of Commerce regulation that interprets the Bayh-Dole Act (incorrectly, in our view and that of the National Academy of Sciences' recent report on Bayh-Dole, see

NAS Report 2010) as placing an affirmative prohibition on *any* release of information from iEdison. The Government Accountability Office has on occasion investigated completeness of reporting to iEdison, but the last such study appears to date from 1999.

The lack of recent empirical data is particularly worrisome because the 1999 GAO report found that, out of a sample of 633 medically related patents issued to 12 academic grantees, 143 had most likely arisen from NIH funding but had neither been reported to iEdison nor contained government interest statements. The National Academy of Sciences has also noted deficiencies with iEdison, suggesting that "without a reasonably complete list of government-sponsored inventions, effective oversight is impossible" (NAS 2010, p. 65).

The need for oversight is far from just symbolic. As universities with large numbers of unlicensed patents come under financial pressure, they may seek to monetize these patents through assertions against those who have commercialized independently (Ledford 2013). Particularly in cases where universities are dealing with patent assertion entities (PAEs), oversight is necessary to ensure that commercialization, not revenue maximization, remains the priority. Even outside the PAE context, government march-in could be an important tool to prevent academic patents from hindering downstream research and development (Rai and Eisenberg 2003). Some scholars (Arno and Davis 2001) view march-in as a tool to help control drug costs, though there is disagreement regarding whether the framers of Bayh-Dole intended it for these reasons (Herder 2008). Health advocacy groups such as Knowledge Ecology International have tried to use march-in to promote access to medicines, though the NIH has thus far denied all such requests.

In addition to march-in, recoupment policies—which would return royalties from profitable drugs to federal coffers—would also require compliance with reporting obligations. Recoupment was originally built into Bayh-Dole to defuse criticism that allowing for patents on public research would lead to "profiteering" at the expense of the public interest (Herder 2008). While it was dropped from the final legislation "in response to concerns that the process for determining repayment was threatening to cause an impasse in deliberation" (DHHS 2001), recoupment provisions occasionally resurface (Korn and Henig 2004), most recently from the NIH director (Sampat and Lichtenberg 2011). Putting aside the question of whether recoupment is desirable, in order for it to be feasible, public agencies would require access to reliable information on patents and products resulting from the research they fund.

In sections 3 and 4 below, we present new empirical evidence on compliance with reporting of government interests in patents and disclosure of publicly funded patents to grant-making agencies.

3. Large-Scale Quantitative Evidence on Compliance

Evidence from Granted Academic Patents, 1977–2007

First, we collected data on all "academic" patents issued between 1977 and 2007, with "academic" patents defined using the Azoulay-Sampat concordance (Azoulay, Michigan, and Sampat 2007). Since we are interested in compliance by extramural researchers, we drop any patents directly assigned to a federal agency (e.g., the Department of Health and Human Services). We focus on the subset of these academic patents in biomedical patent classes (USPTO classes 435, 514, 424, 530, 536, and 600), based on the patent class-field concordance developed by the National Bureau of Economic Research (Jaffe and Trajtenberg 2005). As many authors have noted, biomedical patents are a large share of total academic patents. While academic patents over this period spanned 394 patent classes, 42% of all academic patents map to these six biomedical patent classes.

For each of these 26,943 academic biomedical patents, we collected data on "government interest statements" from the front pages. About 43% of academic biomedical patents have a government interest statement. Figure 3.1 shows this share has generally been increasing over time, and since the early 1990s it has grown particularly rapidly. Over this same period of time, the federal government share of total life sciences research funding has been decreasing, as demonstrated in figure 3.2. Since there is no reason to believe the propensity to patent government-funded (as opposed to, say, institutional or industry-funded) academic biomedical research at universities has increased over this period, we read these trends as reflecting substantial nondisclosure of government interest in the early years. A 43% overall incidence of government interest statements also provides prima facie evidence of underdisclosure, since upward of 60% of academic life sciences research is publicly funded.

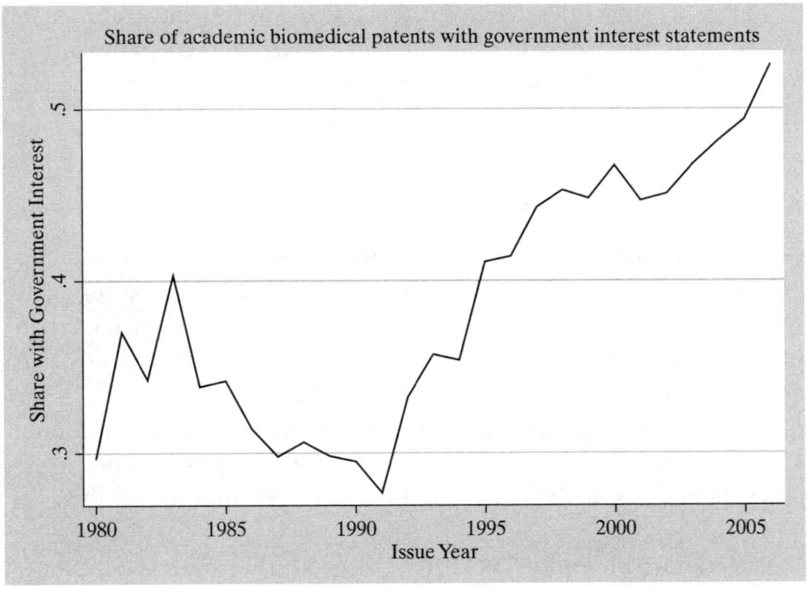

FIGURE 3.1. Share of academic biomedical patents with government interest statements.

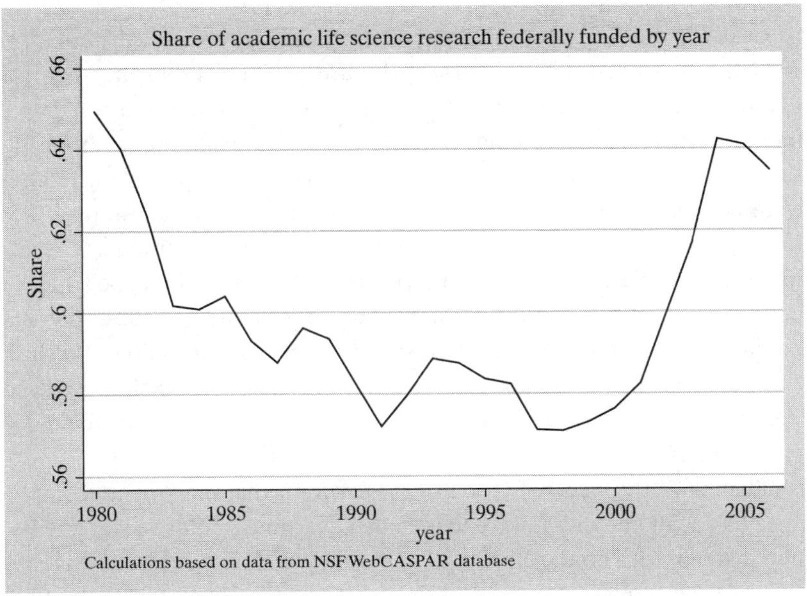

FIGURE 3.2. Share of academic life science research federally funded by year.

Figure 3.3 shows the share of patents with government interest statements for universities with over 75 patents issued to them between 2000 and 2006. These data show considerable variation, even among the largest patentees.

Evidence from iEdison/RePORTER Data

As another window on the questions, we looked at all NIH-funded patents disclosed to the U.S. government. As noted earlier, external researchers do not have access to iEdison and, indeed, even underlying information on patents reported to agencies has not historically been available to the public. The NIH RePORTER database, unveiled in 2010 to provide information on NIH grants in general, now includes this information. However, the data are limited to what is disclosed to the agency and kept in iEdison, as the RePORTER website cautions:

> Patent information in RePORTER is incomplete. The patents in RePORTER come from the iEdison database. Not all recipients of NIH funding are compliant with the iEdison reporting requirements, particularly after their NIH support has ended.

Overall, the RePORTER database includes 11,576 patents, emanating from 8,450 distinct NIH grants. Figure 3.4 shows that the share of patents included in RePORTER that have a government interest statement has increased over time, with the inflection occurring at about the same time as was seen in the share of total biomedical patents with government interest statements, above. In figures 3.1 and 3.2 above, factors other than noncompliance—primarily private or nonfederal funding of research—could explain a reporting rate of less than 100%. Here however, in figure 3.3, this is not possible. By definition, these patents ought to include government interest statements. But 30% do not. While compliance is markedly better over time, with over 80% including these statements in the most recent years studied, noncompliance for the older patents could pose legal problems going forward.

That said, the Federal Circuit case law on sanctions for noncompliance has dealt not with failure to include government interest statements, but, instead, with nondisclosure to the agency. This is unknown, since we cannot (in a large sample) determine which of the academic biomedical

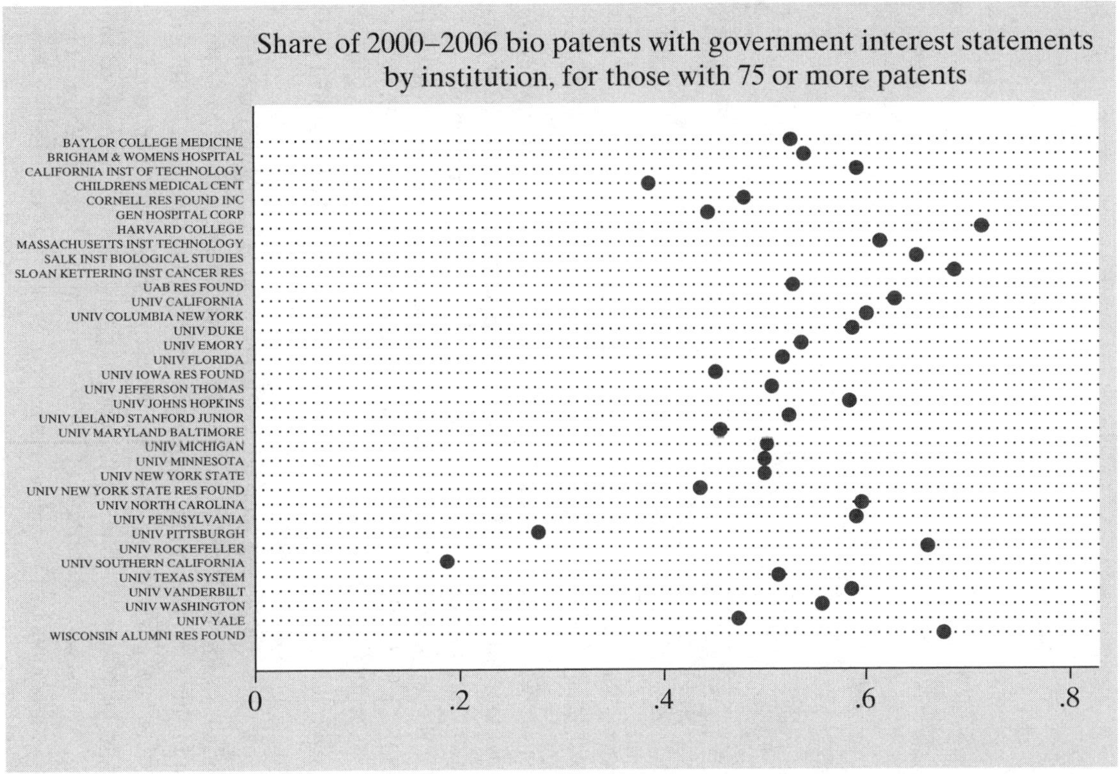

FIGURE 3.3. Share of 2000–2006 bio patents with government interest statements by institution.

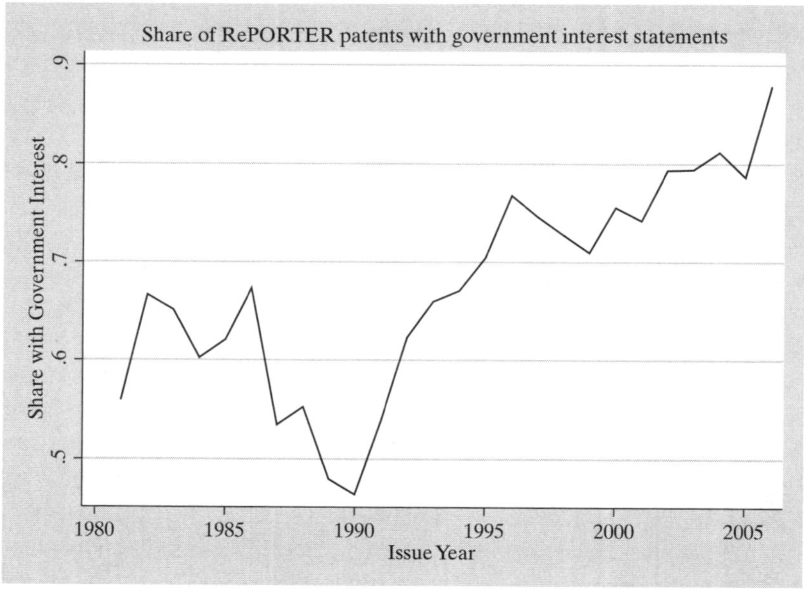

FIGURE 3.4. Share of RePORTER patents with government interest statements.

patents ought to have been disclosed but were not. Table 3.1 below shows a cross-tabulation of whether the academic biomedical patents have government interest statements and are disclosed to the agency, or not.

Of the 26,943 academic biomedical patents, only 30% are included in RePORTER. Of those with government interest statements, about half were in RePORTER; of the rest, only 13% were. Perhaps most noteworthy, over half of these patents were neither in RePORTER nor had government interest statements.

How does the picture look more recently? Table 3.2 shows the analogous figures for academic biomedical patents issued in 2006. Note that, even in this year, the modal category is patents that are in RePORTER and that lack government interest statements. However, there does appear to be better compliance in reporting government interests to the PTO. Almost 90% of RePORTER-listed patents include government interest statements by the end of this period. Somewhat surprisingly, however, the share of patents with government interest statements that are listed in RePORTER is slightly lower than for the overall sample (37%, rather than 50% overall). This discrepancy may reflect lags in updating

TABLE 3.1. **Cross-tabulation of government interest statements versus RePORTER.**

	No Government Interest Statement	Has Government Interest Statement	Total
Not in RePORTER	13,486	5,756	19,242
In RePORTER	1,942	5,759	7,701
Total	15,428	11,515	26,943

TABLE 3.2. **Cross-tabulation of government interest statements versus RePORTER (patents issued in 2006 only).**

	No Government Interest Statement	Has Government Interest Statement	Total
Not in RePORTER	743	541	1,284
In RePORTER	39	326	365
Total	782	867	1,649

the RePORTER data, or including information from iEdison in the RePORTER database.

Overall, the data reported above are suggestive of some noncompliance. Below we examine these issues qualitatively for a sample of patents that are disproportionately important for health, and in debates about recoupment, march-in, and transparency—those associated with marketed drugs.

4. Some Evidence from FDA Approved Drugs

Quantitative Overview

Sampat and Lichtenberg (2011) used data on new molecular entities approved by the FDA between 1988 and 2005 to examine the roles of the public and private sectors in pharmaceutical innovation, including for purposes of assessing how large the scope was for recoupment, march-in, and other policies that leverage government interest to influence drug prices. That study examined patents on these drugs in the FDA *Orange Book*, and also collected information on the government role in these patents: whether they had government interest statements or whether they were assigned to a funding agency (e.g., the NIH). For the purposes of this paper, we will refer to these as "government" patents. Using these data, the authors found that 9% of the drugs approved over this period had government patents. (The percent was higher for drugs that received priority review by the FDA, arguably the most clinically important ones: 17.4%, compared to 3.1% of non-priority review drugs.)

In view of concerns about potential nondisclosure of government interests, the study also looked at drugs with "academic" patents (including those assigned to federal agencies as well as to universities and nonprofits) to assess the public sector role. These data show that 13% of the drugs had an academic patent, including 6% of standard-review drugs and 22% of priority-review drugs.[1]

The difference between the figures for "academic" and "government" patents suggests potential nondisclosure, since most academic research is federally funded. To examine this further, we focus on "academic" patents that are not also "government" patents (which again, means they don't have government interest statements or aren't assigned to a government agency).

Table 3.3 below (reproducing information from footnote 44 of the Sampat and Lichtenberg paper) shows that most of the drugs have neither "government" nor "academic" patents. Of the nonacademic patents, only a tiny share (2%) have government interest statements, as might be expected. However, of the 48 drugs with academic patents, a substantial minority (21, or 44%) had no government interest statements in their patents.

In addition to these drugs (where none of the academic patents acknowledged government support) there were seven others where some of the academic patents acknowledged government support, but others

TABLE 3.3. **Cross-tabulation drugs with at least one "academic" patent versus drugs with at least one "government-funded" patent, based on all new molecular entities approved between 1988 and 2005.**

	No Government Funded Patents	At Least One Government Funded Patent	Total
No Patents with an Academic / Public Sector Assignee	324	7	331
At Least One Patent with an Academic / Public Sector Assignee	21	27	48
Total	345	34	379

did not. Collectively, there are 43 patents associated with these drugs. Of these, seven patents were listed in the NIH RePORTER database, so there was disclosure even if no government interest statements were included.

We focused on the remaining 36 patents (associated with 22 drugs) to assess nondisclosure qualitatively. In the next section, we present findings from these case studies.

Case Studies

Under the Bayh-Dole Act, an invention is a "subject invention" governed by reporting obligations if it was "conceived of or first actually reduced to practice in the performance of work under a funding agreement." In our analysis, we used a conservative definition of "subject invention." For example, when a patent had many inventors and only a few of the inventors appeared to have been supported by federal funding during the time of the relevant research, we deemed the situation "unclear." We also deemed "unclear" the situation where the federally funded research appeared to cover compounds closely related to the patented drug but we could not resolve the question of precise overlap. Even under this conservative definition, it appears that federal funding was not appropriately reported for at least 15 patents (associated with eight drugs). In addition, for five patents (associated with five drugs), federal funding was clearly involved but the applicability of the Bayh-Dole reporting requirement was unclear.

SOMAVERT. An example of a drug that appears to have emerged from federal funding is Somavert (generic name pegvisomant). Somavert is used for the treatment of acromegaly. It is associated with four patents, two for compounds (issued in 1994 and 1997) and two for treatment methods using those compounds (1999 and 2003). The inventors on the patents are John J. Kopchick and Wen Y. Chen, and the patents are assigned to Ohio University.

The patents at issue are all related to antagonists of vertebrate growth hormones obtained by mutation of the third alpha helix of such proteins. Meanwhile, a *Journal of Biological Chemistry* paper published in 1994 that includes Kopchick as the senior author and Chen as the first author specifically discusses "functional antagonists" to growth hormone activity obtained through alteration of the third alpha helix (Chen et al.

1994). This 1994 paper acknowledges support from the NIH and USDA, as does a similar 1995 paper. In 2008, worldwide sales of Somavert were estimated at $250 million.

MERETEK UBT KIT. The patent in question claims a breath test for ulcer bacteria that employs a plastic bag and needle assembly. The inventors on the patent are Antone Opekun and Peter Klein, both of the Baylor College of Medicine, and the patent issued in 1992. Several years earlier, both inventors had published a paper in *Lancet* entitled "Campylobacter pylori detected noninvasively by the 13C-urea breath test." This paper acknowledges support from the NIH and USDA.

ELMIRON. The patent at issue in this case claims a method of treating bacterial infections involving the urinary tract and bladder by irrigating the bladder with a solution containing Elmiron (generic name sodium pentosanpolysulfate). The patent was originally filed in 1982 and ultimately issued (after a series of continuations) in 1993. Lowell Parsons of the University of California, the inventor on the patent, received support from the U.S. government (primarily the Veterans' Administration) from 1979 through 1992. This support was for purposes of working on "antibacterial defenses of the urinary bladder." In 1982, Parsons authored an article entitled "Prevention of urinary tract infection by the exogenous glycosaminoglycan sodium pentosanpolysufate" that acknowledged research support from the U.S. government.

TRISENOX. Trisenox (generic name arsenic trioxide) is associated with four patents that all claim (in slightly different ways) a method for administering a therapeutically effective dose of arsenic trioxide for purposes of treating acute and chronic leukemia. An application associated with the first of these was filed in November 1997 and ultimately issued in 2004. The three other patents were filed in 2004 and ultimately issued in 2005. Raymond P. Warrell, Pier Paolo Pandolfi, and Janice Gabrilove are the inventors of Memorial Sloan-Kettering Cancer Center are the inventors. In 1997 and 1998, Warrell received three grants from the National Cancer Institute to study the use of arsenic trioxide in leukemia.

SENSIPAR. Sensipar (generic name cincalcet hydrochloride) is associated with two patents claiming methods for treating hyperparathyroidism by modulating the activity of a calcium receptor located on parathyroid cells. These patents were filed in 1994 and 1995 and issued in 2000. The inventors on the patents are Edward Nemeth, Bradford van Wagenen, Manuel Balandrin, Eric G. DelMar, and Scott Moe, and the patent is assigned to Brigham and Women's Hospital and NPS Pharmaceuticals.

From 1992 to 1993, Nemeth held an NIH grant entitled "Characterization of the Parathyroid Cell Ca2+ Receptor." The grant's abstract states that the research will test organic compounds for activity on the calcium receptor, thus "enabling a drug development program aimed at the discovery and design of drug candidates for the treatment of hyperparathyroidism." Additionally, in 1998, Nemeth, Van Wagenen, DelMar, and Balandrin coauthored a PNAS paper entitled "Calcimimetics with Potent and Selective Activity on the Parathryoid Calcium Receptor." This paper acknowledged that the underlying research had been funded by the NIH grant mentioned above.

In addition to these five drugs where there was no disclosure of government funding, three drugs were associated with one patent that was disclosed and one patent that was not.

FUZEON. The prominent anti-AIDS drug Fuzeon (generic name enfuvitide) is associated with a patent filed in 1995 and issued in 2000. The inventors on the patents are Dani Bolognesi, Thomas Matthews, and Carl Wild, and it is assigned to Duke University. The patent covers various forms of the synthetic peptide DP-178. Bolognesi and Matthews are also coauthors of a 1995 *Journal of Virology* paper that explores the use of myriad forms of the synthetic DP-178 peptide. The paper acknowledges that direct NIH support for Matthews as well as NIH support for a Center for AIDS Research with which Bolognesi was then affiliated. Notably, the parent application of which the patent is a divisional does contain a government interest statement. Thus it seems fairly clear that the divisional patent itself should have a government interest statement.

THRYOGEN. The patent in question, which ultimately issued in 2002, emerged at the end of a continuation chain originally filed in 1985. The patent claims a thyroid stimulating hormone (TSH) beta chain produced by a cell transformed with a vector comprising a DNA sequence encoding the beta unit of TSH. The inventors on the patent are Ione Kourides and Graham Whitfield of the Memorial Sloan-Kettering Institute for Cancer Research. From 1985 to 1989, Kourides received support from the NIH to study the "Regulation of Alpha and Beta Subunits of TSH."

In 1988, Kourides and Whitfield, together with two others, published an article entitled "The Human Thyrotropin Beta-Subunit Gene Differs in 5' Structure From Murine TSH-Beta Genes." The article acknowledged support from the NIH. Moreover, another Thryogen patent that emerged from the same continuation chain does have a statement of government support. Thus the omission of a statement in this case does not appear justifiable.

PARAPLATIN. Two patents, a compound patent and a method patent, are associated with this anticancer compound. Interestingly, although the pre-Bayh-Dole compound patent has a statement of government interest, and this patent also discusses the use of the compound in cancer treatment, the subsequent method patent that specifically claims treating cancer lacks a statement of government interest.

In addition to these patents that appear to have emerged quite directly from government support, we believe the following five drugs (and associated five patents) pose close questions.

ANGIOMAX. Angiomax (generic name bivalirduin) is a thrombin inhibitor used as anticlotting therapy in patients undergoing coronary angioplasty. The inventors on the relevant patent are John Marganore, John Fenton, and Toni Kline. The patent was applied for in 1990 and issued in 1993. The three inventors are also coauthors of a paper published in 1992 entitled "Structure of the Hirulog 3-Thrombin Complex and Nature of the S' Subsites of Substrates and Inhibitors." Although the patent does not specifically identify Hirulog 3, it discusses hirulogs generally as well as other thrombin inhibitors. In this article, Fenton, a professor at Albany Medical College, acknowledges NIH support. Fenton also received an NIH grant entitled "Human Thrombins" for fiscal years 1985 through 1994 inclusive.

GLEEVEC. Gleevec (generic name imatinib mesylate) is a prominent cancer drug that had worldwide sales of $3.9 billion in 2010. The "academic" patent associated with Gleevec claims a method of treating gastrointestinal stromal tumors (GIST) using imatinib mesylate. (The compound itself is claimed in an older composition of matter patent solely owned by Novartis.) The international (PCT) application for the academic patent was filed in 2001, and the U.S. patent issued in 2005. The patent is assigned to Novartis, Dana-Farber Cancer Institute, and Oregon Health and Science University.

The method patent has ten inventors. Three of these inventors, George Demetri, Michael Heinrich, and Brian Druker, are also coauthors of a 2002 *New England Journal of Medicine* paper entitled "Efficacy and Safety of Imatinib Mesylate in the Advanced Gastrointestinal Stromal Tumors." While this paper acknowledges federal funding for Dr. Heinrich, it acknowledges only private funding for Dr. Demetri (Novartis Oncology and several foundations), and does not acknowledge any funding for Dr. Druker.

Because we did not have evidence of a major federal role in contributing to the conception or reduction to practice of the Gleevec method patent, we placed this patent in the "unclear" category.

ETHYOL. The academic patent in this case claims methods for using Ethyol (generic name amifostine) and related chemicals to treat neurotoxicity and nephrotoxicity associated with the administration of chemotherapeutic agents. The inventors on the patent are Martin Stogniew, David Alberts, and Edward Kaplan, and the patent is assigned to U.S. Bioscience and the University of Arizona. The patent was filed in 1997 and issued in 1999.

Alberts is a prominent cancer researcher and is the named principal investigator on dozens of NIH grants over the last three decades. Several 1996 papers on which he is the first author discuss the application of amifostine in cancer treatment, and one of these papers (titled "WR-1065, the Active Metabolite of Amifostine (Ethyol) Does not Inhibit the Cytotoxic Effects of a Broad Range of Standard Anticancer Drugs Against Human Ovarian and Breast Cancer Cells") acknowledges NIH support. This paper also mentions Ethyol's protective effects on nerve and kidney tissues. However, because no specific paper that we could find both acknowledges NIH support and discusses at length the use of amifostine to treat nerve and kidney tissue, we deemed this case "unclear."

ALIMTA. The "academic" patent on the anticancer drug Alimta is part of a continuation chain originally filed in 1989. The patent was issued in

1994. The sole inventor on the patent is Edward Taylor, and the patent is assigned to Princeton University. Alimta has annual sales in excess of $1 billion.

In a 2003 paper describing a new method of synthesizing Alimta, Taylor mentions that Alimta was first described in a 1992 publication in the *Journal of Medicinal Chemistry*. Taylor was one of eight coauthors on that paper, and the acknowledgments claimed support from Eli Lilly and from NIH grant CA36054. Further research in RePORTER indicates that grant CA36054 was to coauthor Richard G. Moran from the University of Southern California.

Taylor had several grants from the NIH for fiscal years 1986–1991. The abstract for one of Taylor's grants, titled "Design and Synthesis of DDATHF Analogs as Antitumor Agents" (grant number 5R01CA042367–06), states: "We plan to probe the structural requirements for antiproliferative activity by synthesizing a number of carefully selected analogs which differ from DDATHF in such features as the nature of the left-hand heterocyclic ring, deletion of the chiral center at C-6, and the nature, flexibility and dimensions of the middle bridge region. Of primary importance in our synthetic studies will be the development of chiral synthetic procedures in order to simplify the severe synthesis problems associated with the presence of multiple chiral centers in our target analogs" (emphasis added).

The 1992 *Journal of Medicinal Chemistry* paper—which in Taylor's own words was written when "this compound was initially prepared"—is titled "A dideazatetrahydrofolate analogue *lacking a chiral center at C-6*, N-[4-[2-(2-amino-3,4-dihydro-4-oxo-7H-pyrrolo[2,3-d]pyrimidin-5-yl) ethyl]benzoyl]-L-glutamic acid, is an inhibitor of thymidylate synthase" (emphasis added). The paper also describes alterations in the bridge region of the DDATHF molecule. Both of these alterations were stated goals of Taylor's NIH grant, though this grant was not claimed in the paper and the paper is not linked to the grant on the NIH RePORTER.

We placed this patent in the "unclear" category largely because the chemistry is very complex and we do not know whether someone skilled in the art would view the NIH grant, the *Journal of Medicinal Chemistry* paper, and the patent claim as covering the same molecules.

TARGRETIN (BEXAROTENE). The "academic" patent in question, filed in 1992, and issued in 1995, encompasses "bridged bicyclic aromatic compounds and their use in modulating gene expression of retinoid receptors." The inventors are Marcia I. Dawson, James F. Cameron, Peter D. Hobbs, Ling Jong, Magnus Pfahl, Xioakun Zhang, and Jurgen Lehmann. The patent was assigned to SRI International and to La Jolla Cancer Research Foundation

Dawson, Hobbs, Jong, Cameron, and Pfahl were also coauthors on a 1994 paper in the journal *Nature* entitled "A New Class of Retinoids with Selective Inhibition of AP-1 Inhibits Proliferation." In the acknowledgements section, Pfahl and Dawson claim grants from the NCI and NIH. Several of the compounds tested in this paper appear extremely similar to ones claimed, preferred, and demonstrated in the patent.

We placed this patent in the "unclear" category largely because the chemistry is very complex and we do not know whether someone skilled in the art would view the *Nature* paper and the Targretin patent as covering the same molecules.

Our case studies thus reveal some examples of possible (or even relatively clear) noncompliance with reporting requirements. Notably, we did not include in our case studies cases where the federally funded research in question formed only the foundation for more applied research. Thus, for example, we did not include situations where the federally funded research elucidated a biochemical pathway or target, and a patent subsequently claimed a compound that modulated the pathway or target. So our case studies actually understate the degree to which federal funding was a "but for" cause of the drug patents we studied.

Conclusions

Bayh-Dole was not only about promoting university technology transfer, but also about protecting the public interest. The various compliance and reporting provisions—including the requirements to disclose government interests in federally funded patents and to report these patents to funding agencies—were the main ways in which the framers of Bayh-Dole aimed to protect taxpayer interests. That policy issues like march-in and recoupment continue to be discussed thirty years later suggests these public interest aims of Bayh-Dole continue to be salient.

Commenting on the ultimate desirability of march-in or recoupment, either generally or in particular cases, is beyond the scope of this paper. Rather, our main aim is to provide evidence on compliance with the Bayh-Dole reporting requirements. As the sponsors of Bayh-Dole explicitly recognized, information that emerges from reporting requirements is the necessary substrate on which any discussion of march-in or recoupment must be based.

Using data from government interest sections of patents and new information on patent disclosures to the NIH, we found evidence suggestive of nondisclosure. Our qualitative analyses showed that even in the context of important patents—those associated with marketed drugs—there is apparent noncompliance for 15 patents (associated with eight drugs) and possible issues for another five drugs.

To be sure, some of our quantitative evidence is suggestive of trends toward better disclosure over time. This may reflect learning curves in patent management and technology transfer activities (Mowery et al. 2004). Nonetheless, even past misdeeds may be problematic, to the extent that they understate the government roles in private sector innovation, and may even compromise validity of underlying patents.

Going forward, as the activities of patent aggregators receive attention from Congress and the Federal Trade Commission, accurate information of whether federally funded academic patents are an importance source of supply for aggregators will be important. Such information will be particularly important if Congress continues, as it has in the past (Rai and Sampat 2012), to give patents that originate in universities a privileged position in reform to curb patent-based litigation.

Note

1. In more comprehensive analyses, Stevens et al. (2011) performed similar analyses, but supplemented data on Orange Book patents with information from other sources, including AUTM data drug development. Focusing on a slightly later timeframe, this study found 64 new molecular entities with a public sector patent (compared to 48 above). However this study found similar magnitudes on the roles of the public sector in drug development as Sampat/Lichtenberg, with 20% of priority review drugs and 6% of standard review drugs developed at public sector research institutions. (Another difference is that Stevens et al. also looked at non-NMEs; see also Sampat 2007.) Since Stevens and colleagues did not rely solely on the Orange Book, the Stevens analysis also provides information on biotechnology drugs and vaccines, which are underrepresented in the Sampat-Lichtenberg sample. Many of these drugs are filed as biological license applications (BLAs) rather than new drug applications (NDAs) and not subject to Orange Book listing requirements. Based on data provided to us by Stevens

and his coauthors, about half of the BLAs in their sample that stem from public sector research institutions either have government interest statements or were assigned to a government agency.

References

Arno, P. S., and M. H. Davis. 2001. "Why Don't We Enforce Existing Price Controls? The Unrecognized and Unenforced Reasonable Pricing Requirements Imposed Upon Patents Deriving in Whole or in Part from Federally-Funded Research." *Tulane Law Review*.

Azoulay, P., R. Michigan, and B. N. Sampat. 2007. "The Anatomy of Medical School Patenting." *New England Journal of Medicine* 357 (20): 2049–56.

Chen, W. Y., et al. 1994. "In vitro and in vivo studies of the antagonistic effects of human growth hormone analogs." *Journal of Biological Chemistry* 269: 15892–897.

Department of Health and Human Services (DHHS). 2001. "NIH Response to the Conference Report Request for a Plan to Ensure Taxpayers' Interests Are Protected." Washington, D.C.

Korn, D., and S. Heinig. 2004. "Recoupment Efforts Threaten Federal Research." *Issues in Sci. and Tech.* 20 (4).

Jaffe, A. B., and M. Trajtenberg. 2005. *Patents, Citations, and Innovations: A Window on the Knowledge Economy*. Cambridge, MA: MIT Press.

Herder, M. 2008. "Asking for Money Back: Chilling Commercialization or Recouping Public Trust in the Context of Stem Cell Research?" *Colum. Sci. and Tech. L. Review* 9 (203).

Ledford, H. 2013. "Universities Struggle to Make Patents Pay." *Nature* 501: 471–72.

Mowery, D. C., et al. 2004. *Ivory Tower and Industrial Innovation: University Industry Technology Transfer Before and After Bayh-Dole*. Palo Alto, CA: Stanford University Press.

National Academy of Sciences (NAS). 2010. *Managing University Intellectual Property in the Public Interest*. Washington, D.C.

Rai, A., and R. Eisenberg. 2003. "Bayh-Dole Reform and the Progress of Biomedicine." *Law and Contemporary Problems* 66: 289–314.

Rai, A. K., and B. Sampat. 2012. "Accountability in patenting of federally funded research." *Nature Biotechnology* 30: 953–56.

Sampat, B. N. 2009. "Academic Patents and Access to Medicines in Developing Countries." *American Journal of Public Health* 99 (1): 9–17.

Sampat B. N., and F. R. Lichtenberg. 2011. "What are the respective roles of the public and private sectors in pharmaceutical innovation?" *Health Aff* (Millwood) 30 (2): 332–39.

Stevens, A. J., et al. 2011. "The role of public-sector research in the discovery of drugs and vaccines." *N Engl J Med.* 364 (6): 535–41.

U.S. General Accountability Office (GAO). 2003. "Agencies' Rights to Federally Sponsored Biomedical Inventions" GAO-03-536.

U.S. Senate. 1979. Report of the Committee on the Judiciary on S. 414, "University and Small Business Patent Procedures Act" 1979.

CHAPTER FOUR

An Entrepreneur's Guide to the University

Fiona Murray and Julian Kolev

An Entrepreneur's Guide to the University

How can entrepreneurs engage effectively with universities to translate novel advances in science and engineering from idea to impact? What are the legal rules shaping their engagement, what are the local rules and norms that entrepreneurs must understand, and what incentives and expectations guide faculty attitudes toward entrepreneurs? This chapter aims to answer these questions by examining the topic of "technology transfer," not from the traditional perspective of the faculty member or technology transfer officer, but instead from the entrepreneur's point of view.

The role of entrepreneurs and entrepreneurial activities on university campuses in the United States and the rest of the world has been the subject of increasing attention over the past several decades. Long gone is the view that universities can survive as isolated ivory towers producing fundamental knowledge for the betterment of mankind. In its place is a view that universities should serve as a critical pillar in the knowledge economy, generating fundamental ideas and simultaneously contributing to useful products and services. While few dispute this shift in emphasis, it raises the critical question of how to best transition from idea to impact when commercializing universities' science and engineering research. In other words, will ideas that have both scientific merit and immediate economic value—i.e., those that lie within Pasteur's Quadrant (Stokes 1997)—*actually* move beyond the university's walls and into the economy with maximal impact?

Early approaches (as well as scholarly studies) focused on shaping the broad institutional context in which ideas were generated; specifically, they targeted the ownership of and obligations around intellectual property created through university research. More recently, attention has shifted to the ways in which academics in science and engineering engage in the commercialization of their research. In the U.S., funding agencies including the National Science Foundation, the National Institutes of Health, and the Department of Energy have all provided funding and education to focus the attention of university faculty on moving their research through the technology transfer process. Scholars have also sought to understand the characteristics of those faculty members who do engage in commercialization, including the assessment of the roles of gender, rank, tenure, and productivity (Murray 2004; Azoulay, Ding, and Stuart 2007; Agrawal and Henderson 2002). Nonetheless, the critical role of the entrepreneur in the *process* of moving from "idea to impact" is missing.

This chapter places the entrepreneur *front and center* in the process of commercialization and examines the most effective ways to take on this role. This is the perspective that entrepreneurs outside the university must take as they seek to step inside the corridors of academia and find ideas to transfer and transform. However, it is also the perspective that internal entrepreneurs—faculty, science and engineering students, or those in medical fields—should follow as they shift from a purely technical role to one that involves active entrepreneurship. Specifically, they must shift their primary focus to the task of systematically transforming nascent ideas into meaningful products and ventures. For both internal and external entrepreneurs, there are three distinct "levels" of technology transfer to explore and comprehend:

- First, the *national level*: the institutional rules and legal environment for commercialization; this is the aspect of technology transfer most often described in previous literature, and is centered on the 1980 Bayh-Dole Act (henceforth, BD).
- Second, the *local level*: the local interpretation of the BD Act, including rules established by a given university for determining how ideas are typically accessed by entrepreneurs, how they are licensed, the contractual structure

of licenses (between the university and firm), and the role of faculty in the process.

- Third, and perhaps most important, is the *individual level*: the social networks and norms as well as the incentives for the different actors involved in commercialization, focusing primarily on the structure of the relationship between entrepreneurs and researchers that must be established for effective commercialization.

Without an understanding of each of these different levels of analysis, an entrepreneur is likely to become disoriented within the system, make mistakes or miss opportunities, or at best, slow down the commercialization and reduce the economic and social impact of a potentially promising idea.

This entrepreneur's guide to the university is designed to overcome some of these challenges by laying out each of the three levels described above from the vantage point of the entrepreneur. The analysis of the levels is based on both systematic research of the university and its role in commercialization, as well as on anecdotal evidence from MIT and beyond gathered by the first author over the past dozen years. Its focus is intentionally U.S.-specific, not only because this is the domain of experience of the authors but also because the legal rules are shaped at the national level, and local interpretations, while exhibiting significant variation, also conform to a set of well-established American practices. Lastly, most (although not all) of the statistical analysis and qualitative studies of technology transfer have focused on U.S. universities (see the chapter by Siegel and Wright in this volume for a comprehensive review of this evidence). Comparative analysis across nations remains a topic for future research.

The discussion begins with a commercialization "story" which allows us to introduce the cast of characters typically involved in identifying an interesting scientific idea, determining whether it has some commercial potential, and moving it out of the university toward economic and social impact. In the remaining three sections, this narrative is retold and generalized according to the three levels of analysis outlined above: national institutions, local rules, and individual networks and incentives. We conclude with specific lessons for entrepreneurs.

The Commercialization Stage

If commercialization were theater, the cast of characters would be long, their roles complex, and their motives and interactions entangled. For the entrepreneur this is both the excitement and the frustration of commercialization. By contrast, the theater of more traditional scientific work is relatively straightforward: a faculty member at the university raises funding for his or her research from a variety of interested funders. Typically (about 60% of the time) in the United States, the funder is the federal government, which funds over $30 billion in university research each year. Other major funders include state government agencies, industry, not-for-profit foundations, and the university itself. The negotiations between researchers and those who fund them are subtle (see Gans and Murray 2013), but typically focus on the scope of the research and its disclosure—patenting and future rights to the patents (more of this later). The other critical characters are the students in the research laboratory—undergraduates gaining research experience, masters students completing a short thesis, and PhDs writing a doctoral dissertation while focusing on research, publications, and a future career (in academia or elsewhere). Large laboratories often have one or more postdoctoral research scientists—individuals who have completed their studies but still work in a lab in order to learn new methods, develop an independent reputation, and explore potential career paths (again from academia to industry). Once the lab has generated novel research results, the faculty and various students and scientists will focus on publishing and in some cases patenting—more on this below. For results with intriguing potential for usefulness, the stage now expands and the play shifts to commercialization (while much of the lab's time and attention refocuses to the next research project).

With commercial interests come new characters: a technology licensing officer to help determine if the idea should be patented (and if so, to take the lab and their documents to the patent lawyer). An entrepreneur is also needed—someone willing to drive the idea out of the laboratory into the outside world. Who should play this role? As we will see, many people can—an outside entrepreneur who meets one of the students in a business plan competition or a class, the postdoc taking on the dual role of researcher-entrepreneur, occasionally the faculty member herself. Take, for example, the story of Matthias, a mid-career MBA student with a successful entrepreneurial track record. He found an idea—for power electronics in cell phone base-station towers—in the classroom and worked for a semester to explore its potential with no formal relationship with the lab save an educational one. At the end of the semester, he spent time with Professor Joel Dawson discussing the real potential of the idea, and the commitment of the team to commercialization. After each side vetted the other, Matthias, looking ahead, turned to worrying about how to actually access the idea—how to license it from the university. He wondered, was this the right moment? Was the technology ready to stand alone outside the lab? Would someone from the lab team come along to ensure that things work? The team took an 18-month hiatus for more research to be done and then initiated licensing talks, founded a venture—*Eta Devices*—and brought still more new characters onto the stage: possible early-stage investors, board members, technical advisors, and corporate partners. The licensing negotiation took several months, with Matthias engaging directly with the officer responsible for the electrical engineering department. At this point, Professor Dawson joined him full-time, leaving the confines of academia, and the story begins to resemble the plot of traditional entrepreneurship. The commercialization process is far from over, but the complex middle world between the traditional lab and the traditional start-up venture has been navigated.[1]

A single story from the many examples of university commercialization emphasizes the complex role of the entrepreneurs navigating the technology transfer ecosystem. Other stories illustrate how external entrepreneurs can also come in and play Matthias's role, or how individuals including the professor himself, a technical PhD student or postdoc, or even the tech transfer officer, might play a role serving as the initial entrepreneur driving commercialization (see Murray 2004). The remainder of this chapter aims to decode the underlying rules, norms, and motivations for the cast of characters in the ecosystem, told from the vantage point of the entrepreneur, whoever he or she may be.

1. National Institutions

The Bayh-Dole Act

Passed by the U.S. Congress in 1980, the Bayh-Dole Act (henceforth referred to as the BD Act) governs the legal framework of technology transfer from universities to private sector entities in cases where research is funded (at least in part) by the federal government in the United States. The Stevenson-Wydler Act (henceforth, SW) was passed in the same year with an emphasis on transferring technology from the U.S. complex of national laboratories. In practice, this includes the vast majority of research performed at top U.S. universities, and therefore determines the structure of intellectual property (IP) control from the perspective of the entrepreneur.[2] Indeed, with federal funding to universities for science and engineering research making up almost 65% of the $61 billion in total R&D expenditures as of 2011,[3] the BD Act covers a significant fraction of university research that entrepreneurs are likely to attempt to access. As laid out in the legislative language, the SW Act was intended "to promote the United States' technological innovation for the achievement of national economic, environmental, and social goals." Similarly, the BD Act states that "it is the policy and objective of the Congress to use the patent system to promote the utilization of inventions arising from federally supported research or development . . . to promote the commercialization and public availability of inventions made in the United States by United States industry and labor."[4] For entrepreneurs engaging with U.S. universities, the BD Act and provisions thereof have become the most important national-level laws and institutions shaping their interactions in the technology transfer process.

The primary effect of the BD Act is to allow universities to maintain control over the rights to their inventions; this provision is accomplished largely through the patenting process. In practice, this means that the university identifies ideas generated by its faculty (and students working with federal funding in faculty laboratories). It then solicits so-called invention disclosures that briefly describe the novel and useful idea, and makes a determination as to whether to file for IP rights over the idea. (Universities are not obligated to file patents on all invention disclosures, but instead make a determination of potential patentability and likely commercial potential). If and when patents are filed, the university owns the IP rights to the inventions of its faculty, while faculty and other participating inventors are listed as inventors on the patent. (This is similar to the practice where corporate researchers serve as inventors but not owners of the IP generated in the course of their employment). This is illustrated in figure 4.1—the image of a patent filed on carbon nanotubes by faculty member John Vander Sande at MIT and assigned to the Massachusetts Institute of Technology. It is worth noting that for entrepreneurs, each university's portfolio of (granted) patents is easily search-

(12) **United States Patent**
Goel et al.

(10) **Patent No.: US 7,641,882 B2**
(45) **Date of Patent: Jan. 5, 2010**

(54) **FULLERENIC STRUCTURES AND SUCH STRUCTURES TETHERED TO CARBON MATERIALS**

(75) Inventors: **Anish Goel**, Washington, DC (US); **Jack B. Howard**, Winchester, MA (US); **John B. Vander Sande**, Newbury, MA (US)

(73) Assignee: **Massachusetts Institute of Technology**, Cambridge, MA (US)

(*) Notice: Subject to any disclaimer, the term of this patent is extended or adjusted under 35 U.S.C. 154(b) by 0 days.

(21) Appl. No.: **10/675,140**

(22) Filed: **Sep. 30, 2003**

(65) **Prior Publication Data**

US 2005/0095191 A1 May 5, 2005

(51) **Int. Cl.**
C01B 31/02 (2006.01)

(52) **U.S. Cl.** .... **423/445 B**; 423/449.1; 423/449.2

(58) **Field of Classification Search** ...... 423/449.1–450, 423/445 B

See application file for complete search history.

(56) **References Cited**

U.S. PATENT DOCUMENTS

5,132,105	A *	7/1992	Remo	423/446
5,648,523	A *	7/1997	Chiang	562/100
5,851,503	A *	12/1998	Mitani et al.	423/445 B
6,358,375	B1 *	3/2002	Schwob	204/173
2002/0122754	A1 *	9/2002	Ryzhkov	422/186.21

FOREIGN PATENT DOCUMENTS

JP 11-140342 * 5/1999

OTHER PUBLICATIONS

Taylor, R. and Walkton, R.M., The Chemistry of Fullerenes, 363 Nature 685 (Jun. 24, 1993).*

Taylor, et al., The chemistry of fullerenes, Nature 1993; 363: 685-693.*

Donnet, et al., Fullerenic carbon in carbon black furnaces, Carbon 2000; 38: 1885-1886.*

Burden, et al., In-situ Fullerene Formation—The Evidence Presented. Carbon 1998; 36(7-8): 1167-1173.*

Burden, et al., In-situ Fullerene Formation—The Evidence Presented [online] American Carbon Society [http://acs.omnibooksonline.com/papers/1997_ii376.pdf].*

Berezkin, Fullerenes as Nuclei of Carbon Black Particles, Physics of the Solid State 2000; 42(3) 580-585.*

Cataldo, The impact of a fullerene-like concept in carbon black science, Carbon 2002; 40: 157-162.*

Donnet, Black and White Fillers and Tire Compound, Rubber Chem. Tech. 1998; 71(3): 323-341.*

(Continued)

Primary Examiner—Stanley S. Silverman
Assistant Examiner—Daniel C. McCracken
(74) *Attorney, Agent, or Firm*—Choate Hall & Stewart LLP; Sam Pasternack

(57) **ABSTRACT**

The fullerenic structures include fullerenes having molecular weights less than that of C_{60} with the exception of C_{36} and fullerenes having molecular weights greater than C_{60}. Examples include fullerenes C_{50}, C_{58}, C_{130}, and C_{176}. Fullerenic structure chemically bonded to a carbon surface is also disclosed along with a method for tethering fullerenes to a carbon material. The method includes adding functionalized fullerene to a liquid suspension containing carbon material, drying the suspension to produce a powder, and heat treating the powder.

18 Claims, 8 Drawing Sheets

FIGURE 4.1. United States Patent No. 7,641,882.

able via the U.S. Patent and Trademark Office search engine using the university name (or the specific name under which the university files its patents).[5]

Obligations on Universities

The Bayh-Dole Act specifies a number of obligations that the university must meet as a consequence of their taking title to inventions and subsequent intellectual property: invention disclosure, federal license, and intellectual property protection.

First, universities must disclose any identified inventions to the federal agency providing funding. While this action lies outside the scope of the entrepreneur's control, it is often valuable to verify that these requirements have been fully satisfied, as there are a number of potential pitfalls. In the case of disclosure requirements, failure to fully disclose the invention can lead the university to forfeit control over the invention, invalidating the entrepreneur's ability to license it, as determined in *Campbell Plastics Engineering and Mfg., Inc. v. Les Brownlee*, 389 F.3d 1243 (Fed. Cir. 2004).

Second, the university must grant the federal government a (nonexclusive, nontransferable) license to the invention. This is generally not a cause for alarm by the entrepreneur, as the federal government is almost certain to be uninvolved in commercialization or competition with private-sector industries (Duecker 1997). However, while commercialization by the government is unlikely, experimental use by the federal government or its agents is permitted under the decision in *Madey v. Duke University*, 307 F.3d 1351 (Fed. Cir. 2002)—"in light of the Bayh-Dole Act . . . use of the patents that has been authorized by the government does not constitute patent infringement." Despite this, there is no precedent for the government or its agents entering into product market competition with an entrepreneur licensing an academic patent under the Bayh-Dole Act.

Third, the university is also responsible for protecting its intellectual property. This can be seen as a benefit, as the entrepreneur will not be directly responsible for legal costs in the event of infringement. Importantly, while universities do usually request reimbursement for patent filing fees in the case of exclusive licenses, the risk of incurring costs defending the patent remains largely in the hands of the university, unless otherwise specified under the licensing contract. To take a prominent example, MIT has defended its IP numerous times in recent years. In 2007, in conjunction with its licensee Biothera, it filed suit against Brazilian company Biorgin for six counts of patent infringement for selling gluco-polysaccharide ingredients in the U.S. for nutritional supplements and functional foods.[6] In 2003, MIT filed suit against Hitachi for possible infringement in a decision that went against the university. In the same year, the university also sued Lockheed Martin for infringement of a 1989 MIT patent issued for an invention by two Lincoln Lab researchers, related to the operation of its two "Inmarsat" satellite communications ground stations.[7] Interestingly, the infringing technology used by Lockheed was bought from an MIT spin-off company, Digital Voice Systems Inc, which had licensed a different set of MIT patents for speech compression.[8] As these examples demonstrate, MIT and other universities have demonstrated a willingness to actively defend their IP when they suspect infringement, even in the face of well-funded corporations or well-connected former spin-offs. For the entrepreneur, this should provide strong assurance that any license agreement will be well-defended against improper competition.

Conditions of Commercialization

While the stipulations described above focus on the role of the university in filing and defending its IP, the Bayh-Dole Act includes several conditions specifically targeting commercialization, which are of central interest to the entrepreneur. The most significant is the ability and requirement of universities to grant licenses (but not outright transfers of title) to their intellectual property, in order to facilitate the commercialization of faculty inventions. Indeed, both BD and SW mandated universities and national laboratories to seek technology transfer as part of their core mission, and to create technology transfer offices to carry out this task. In giving universities the scope to grant such licenses, the Bayh-Dole Act included important stipulations regarding revenue sharing. Specifically, the university must share licensing revenues with the original inventor. Any remaining revenue after expenses must be used to support research and education. This is important to the entrepreneur in two ways. First, the revenue stream to the original inventors provides at least a partial incentive for them to engage with the entrepreneur and contribute to the commercialization and transformation of the idea; this aligns all parties behind the goal of maximal impact. Second, the requirement that remaining revenues go to research allows for the possibility of using the licensing-based revenues to support both the original research lab as well as other labs at the university working to extend the broad research area of the initial invention. While the licensee does not have any control of the direction of research undertaken using this revenue stream or the licensing terms of any subsequent IP, this requirement does offer the potential for a positive-feedback effect between successful commercialization and funding for research.

A second stipulation that pertains to any licensee, but which is a source of particular concern to (and frequent misunderstanding by) entrepreneurs, is the granting of march-in rights. Under the BD Act, there exists a provision for granting "march-in" rights to third parties in the event that commercialization is not proceeding satisfactorily under a university-granted license. In theory, this has the potential to dilute the value of the entrepreneur's license, particularly in cases of exclusivity. In practice, the concerns are limited and few instances can be found to support the view that the average entrepreneur should be nervous of march-in provisions. Owing to the tension between theory and practice, it is worth understanding the provisions in detail, as they define the boundaries between the entrepreneurship encouraged by the BD Act, and the practice of "patent hoarding" that it seeks to avoid. Formally, the BD Act contains four conditions, each of which can lead to the granting of march-in rights:[9]

1. When commercialization of the invention is found to be taking an "unreasonable" amount of time.
2. If there is any danger to health and safety resulting from the existing licensing arrangement (e.g., in the case of essential medical supplies), additional licenses may be granted.
3. If there is a danger that public use of the invention will be interrupted, march-in rights may be invoked as a remedy.
4. Failure to comply with the requirement for domestic manufacturing can lead to march-in rights being granted.

On four occasions, competitors to university patent licensees have filed official petitions with the National Institutes of Health (NIH) for march-in rights, often focusing on the "health and safety" clause of the second condition. In all four cases, the NIH declined to institute march-in rights; however, the details of these decisions provide entrepreneurs with guidance for navigating the technology transfer process under the BD Act. The first petition for march-in rights was filed in 1997 by CellPro Inc.—the first company to reach the product market with an FDA-approved device for stem cell separation. CellPro's petition was aimed at Baxter Healthcare Corp., which held an exclusive license to the underlying technology, based on patents held by Johns Hopkins University. Despite the fact that CellPro was faster than Baxter in reaching market with its device, the NIH ruled that Baxter was *not* taking an unreasonable amount of time in commercializing the invention: Baxter's competing device was undergoing clinical trials. Health and safety were not threatened because Baxter had not used its exclusivity to prevent sales of CellPro's device in the period prior to Baxter's device obtaining approval. This ruling set a precedent whereby march-in rights will not be granted even if an entrepreneur takes more time to reach market than the competition. However, if this occurs, entrepreneurs will be unable to prevent competitors' sales until their own product is ready and has received regulatory approval. Importantly, this applies only if the product satisfies a health or safety need; if this is not the case, the license holder may petition for an injunction against the competitor's sales without opening the door to march-in rights.

The next two petitions were both filed in 2004, under very similar circumstances. The petitions concerned the prescription drugs Norvir and Xalatan. In the case of Norvir, the petition was filed in response to a significant price increase by the patent holder, Abbot Laboratories, which had developed Norvir partly through the use of federal funds. With Xalatan, the march-in petition was filed on the basis that license holder Pfizer, after licensing patents from Columbia University, had set the U.S. price higher than the price in Europe or Canada. In both cases, the NIH based its conclusions on the fact that that Norvir and Xalatan had already been approved by the FDA as safe and effective and were broadly available. Because of this, both cases met the requirements of achieving commercialization in a reasonable amount of time, and satisfying the health and safety needs of the public. As a result, the NIH concluded that march-in rights were not an appropriate remedy for concerns regarding price.[10] For the entrepreneur, these rulings mean that pricing decisions will have no impact on the potential for granting march-in rights. While it is possible that some pricing structures may draw scrutiny from other sources, an exclusive license obtained under the Bayh-Dole Act is unlikely to be affected.

The most recent petition for march-in rights was filed in 2010, and dealt with the prescription drug Fabrazyme. The basis for the petition was a shortage owing to production difficulties by Genzyme, the license holder, as it was commercializing patents held by the Mount Sinai School of Medicine. The shortages led to rationing of the drug and lower-than-optimal treatment levels, leading to significant concerns regarding health and safety. March-in rights were not granted, because of practical limitations on competitors' ability to reach market quickly with an alternative product; however, under different circumstances, the decision could easily have been otherwise. From the ruling, if "a third party has a viable plan to obtain FDA approval . . . during the period in which Genzyme is unable to meet patient demand. . . . NIH will immediately re-consider its decision to exercise its march-in authority." Importantly, it is entirely possible for a competitor to prepare a competing product if a shortage is expected, as clinical trials are protected from patent infringement suits under the Hatch-Waxman safe harbor provision (35 U.S.C. § 271[e]). Of the four march-in petitions, this was the closest the NIH has come to revoking the exclusivity of the license holder. From the entrepreneur's perspective, the ability to meet demand is therefore of great importance in retaining an exclusive license, more so than either reaching market quickly or achieving a low price structure for consumers. This applies most strongly to products like Fabrazyme, which are essential (and frequently the only) treatments meeting the "health and safety" needs of a given patient population. Thus, while the lack of alternative treatments is often considered an advantage from an entrepreneurial perspective, it should also be seen as a significant risk for march-in rights in the event of difficulty meeting demand in areas of medical need.

Selecting Licensees

In selecting potential licensees, the BD Act gives universities considerable leeway. However, it stipulates defined preferences for two aspects of commercialization: domestic manufacturing and small businesses. In specifying a preference for domestic manufacturing, the BD Act notes that if the company holding an *exclusive* license wishes to sell a product in the U.S., it must "substantially" manufacture that product domestically, or obtain a waiver for this requirement. Importantly, this does not bar international businesses or entrepreneurs from taking advantage of university innovations under Bayh-Dole—it simply requires that manufacturing of goods for the domestic market occur domestically. If the entrepreneur seeks to license a patent with the intention of providing a service rather than a commercial product, then this requirement does not apply. Further, a waiver is available if the entrepreneur can show that "reasonable but unsuccessful efforts had been made to find a company that would manufacture the product in the U.S., or that manufacture in the U.S. would not be economically feasible."[11] Because of these details, the domestic manufacturing requirement is relatively limited in scope, and is unlikely to act as a significant impediment to commercialization from the entrepreneur's perspective.

The requirement that universities show a preference for "small businesses"—defined as firms with fewer than 500 employees—does seem to provide an advantage for entrepreneurs, as they have higher priority if they are competing for a license with a larger, better-established company. However, as with the domestic manufacturing requirement, this preference is administered with significant flexibility. The first source of flexibility is that if the technology transfer office *believes* a small business lacks the resources and capabilities to effectively commercialize the invention, the university is then free to select a large corporation as the licensee. Because this is a subjective judgment on the part of the university, it offers significant leeway (and room for local interpretations) in the decision-making process. This underscores a point we will emphasize later in this guide—there is a significant benefit to having strong relationships with both the technology transfer office and the original inventor, as the entrepreneur would gain an advantage in the event of competition for licensing rights.

Taken together, the above descriptions detail the provisions of the Bayh-Dole Act (particularly as it pertains to entrepreneurs), with an emphasis on legislative stipulations, interpretation, and implementation at the national level. While dimensions such as the location of manufacturing and the ability to meet health and safety needs are clearly spelled out in the original legislation and subsequent rulings by federal agencies and the courts, these only cover a small portion of the commercialization process. The numerous dimensions not pinned down by the laws and decisions above offer a wide range of leeway to universities (and indeed to individual faculty) to create local norms and practices. While this also gives the entrepreneur flexibility and choice regarding how he or she effectively manages commercialization, the entrepreneur does not have complete freedom: in addition to abiding by the national-level requirements of the Bayh-Dole Act, it is crucial that entrepreneurs collaborate effectively with both the university and the original inventors. They can best do so by understanding the challenges and the opportunities presented at next two levels of technology transfer: the local norms and practices of the university, and the individual incentives and social networks of faculty and students. We outline these topics in the following sections.

2. Local Norms and Interpretations

In this section, we proceed from the national-level perspective to the local level, in order to highlight salient differences among universities in the ways in which technology is moved into the private sector, toward commercialization. At this level, we focus on three main topics—the typical path of ideas from the lab out into the commercial ecosystem (as well as local variation), typical expectations for the frequency with which this transformation takes place, and how entrepreneurs can gain access to ideas as they pursue commercialization. This includes engaging with technology transfer offices and the details of licensing contracts. By the end of this section, entrepreneurs should have an understanding of how to proceed from an interest in a given field of research to having a licensed technology in their hands. It is worth noting that these topics are rarely the first chronological steps for the entrepreneur to take in the process of technology transfer; rather, we present them as a general orientation before exploring the details of personal connections in section 3.

Technology Transfer Offices

Following the passage of the Bayh-Dole and the Stevenson-Wydler Acts of 1980, all universities were eligible to receive grants for the establishment of what the law described as cooperative research centers. These centers were meant to enhance innovation by improving the cooperation between universities and industry; virtually all major research universities took advantage of these grants. Within these research centers, the point of contact with industry is handled by dedicated technology transfer offices (TTOs). Over time these offices have come to serve a central role in patenting and licensing ideas across the entire research enterprise of the university. Broadly speaking, most TTOs provide technical, business, and legal expertise to facilitate the process of identifying ideas with commercial potential among their university's laboratories; once an idea is identified, they work to move it into the private sector via patents and licenses.

While this overall TTO approach is standard across U.S. universities, there is significant variation among their mission statements, the details of their technology transfer process, and their specific legal and reporting structures; effective technology transfer depends on the entrepreneur's understanding of these local characteristics. For example, MIT's TTO is effectively a department of the institute and reports to the vice president for research with a stated mission: "to benefit the public by moving results of MIT research into societal use via technology licensing, through a process which is consistent with academic principles, demonstrates a concern for the welfare of students and faculty, and conforms to the highest ethical standards."[12] By contrast, the University of Wisconsin office that supports technology transfer—the Wisconsin Alumni Research Foundation (WARF) is a separate, private, nonprofit organization[13]—states its mission differently, emphasizing the connection between licensing and future research: "The mission . . . is to support scientific research at the UW–Madison. WARF accomplishes this by patenting inventions arising from university research, licensing the technologies to companies for commercialization, and returning the licensing income to the UW–Madison to support further scientific endeavor."[14]

Within all TTOs, regardless of their precise structure and mission, ideas and potential inventions follow a well-specified path: identification of a possibly interesting and useful idea, invention disclosure, assessment of potential, patent filing (and later patent grant), identification of possible licensees, and then license option and formal licensing. (It should be noted that formal licensing can come before or after the patent grant). This process runs in parallel with the traditional steps of academic disclosure, with significant interactions between the two. (See figure 4.2 for a simple flow diagram.) In particular, faculty are able to disclose their ideas in both publications and in patents, so long as the patent application is filed no later than one year after publication.[15] This places a burden of rapid patent filing on licensing offices and interested entrepreneurs, particularly when faculty are somewhat tardy in their in-

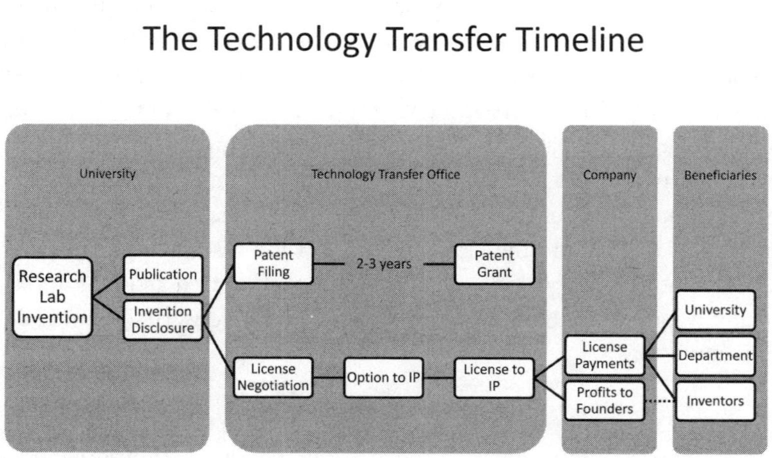

FIGURE 4.2. The Technology Transfer Timeline.

vention disclosures! These simultaneous disclosures of publications and patents have been referred to as "patent-paper pairs" (see Murray 2002 and Murray and Stern 2007 for a more detailed description) and have been the chosen disclosure strategy for a number of prominent scientific ideas:

- The techniques of recombinant DNA were disclosed in patents on the production of recombinant therapeutic proteins and publications providing insights into the rDNA cellular machinery of the cell (Cohen et al. 1973).
- The three-dimensional printing (3DP) breakthroughs analyzed in a paper by Shane (2000) were not only disclosed in patent 5,204,055 (and subsequently licensed to more than eight different parties) but were also disclosed in a publication of *Journal of Engineering for Industry* (Sachs et al. 1990).
- Nobel Prize–winning research on chemical carbon fullerene "Bucky balls" was both patented and published (Guo, Jin, and Smalley 1991).

Identifying Commercial Opportunities

Beyond describing the simple flow of ideas from labs to product markets, the current literature provides few insights into the nuances of the commercialization process, and has even less to offer on the role of entrepreneurs in this flow. Despite this, one can take advantage of recent findings regarding the number of ideas and the rate of attrition at each stage of technology transfer, and the most typical sources of opportunities for entrepreneur-driven commercialization. At this point, it is worth reminding enthusiastic entrepreneurs, faculty, or lab members that not all good ideas necessarily embody commercial potential. Consequently, technology transfer officers and entrepreneurs (as well as faculty and lab teams) have an important role to play in identifying *potential* intellectual property and ideas which hold commercial promise.[16] There are two primary dimensions of the identification process: the specific technological field of the IP, and the characteristics of the researcher or research team behind the innovation. The entrepreneur's search should give significant weight to both of these dimensions, and we describe them in detail below.

(A) IDENTIFYING TECHNOLOGICAL FIELDS. The first dimension that the entrepreneur should evaluate is the traditional commercial potential of the technological field in question; not all fields hold similar potential for real-world usefulness. As Stokes(1997) pointed out in a seminal piece of science policy writing, researchers can undertake projects aiming to solve real-world problems, aiming to extend the frontiers of knowledge, or both.

While many entrepreneurs assume that *all* university ideas may solve useful problems (and thus hold commercial potential), in fact, many more projects are targeted at expanding knowledge frontiers. To illustrate this distinction, one might consider particle physics, cosmology, paleontology, and plate tectonics as examples of fields that seek primarily to expand current knowledge. By contrast, problems geared purely toward solving real-world problems include such inventions as the light bulb or Internet search algorithms. An emphasis on the production of immediately useful knowledge does exist at some universities and institutes—for example, the Fraunhofer Institutes, extension schools at land grant colleges (CITE), or U.S. national laboratories (see Ham and Mowery 1998). Finally, there is a category of projects that are dual in nature—they aim to both expand knowledge and solve existing problems. If we categorize research projects along the above dimensions—namely, scientific merit and economic usefulness—we arrive at a matrix with four quadrants, as illustrated in figure 4.3 below (Stokes 1997; Murray 2002).

Many research paths offer value along one of the above dimensions; however, only the ideas and inventions in the top-left quadrant offer an opportunity to align the interests of academics and entrepreneurs. This region is called Pasteur's Quadrant, after the eponymous French researcher.[17] It is in this quadrant that commercialization offers the greatest benefits, and identifying these opportunities is central to the entrepreneur's strategy in technology transfer. A cursory examination of the ratio of papers to patents (ranging from 10:1 to 100:1) emphasizes the limited supply of ideas in this quadrant.[18] While the ratio of patents to papers argues for the fact that not all research projects likely yield commercial opportunities, an examination of the fields and departments that have yielded patented research and have commercialized research in the past provides a useful starting point for an entrepreneur. For example, an analysis by Ding, Murray, and Stuart (2006) finds that even in the life sciences, where many novel research projects yield patentable useful results, only one in four faculty have filed for IP rights (from a sample of over 5,000 life science Ph.D. faculty at the top 100 leading research

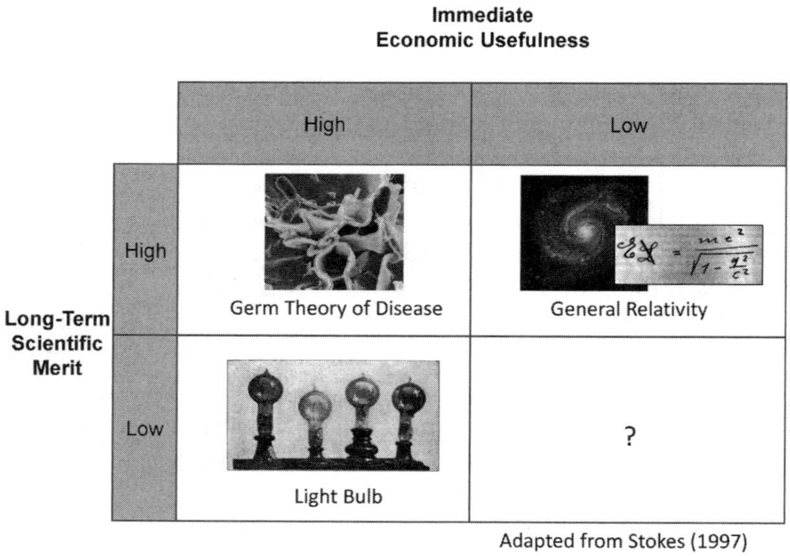

FIGURE 4.3. Pasteur's Quadrant.
Credit: Germ Theory of Disease: Rocky Mountain Laboratories, NIAID, NIH; Cosmic Whirpool: S. Beckwith (STScI) Hubble Heritage Team, (STScI/AURA), ESA, NASA; The E=mc² equation: Einstein's 1912 Manuscript on the Special Theory of Relativity; Light bulb: Elmer Ellsworth Burns (1910), *The Story of Great Inventions*, Harper & Brothers, New York, p.123, fig.59.

universities). Similarly, Agrawal and Henderson (2002) found that in the late 1990s, less than 20% of mechanical engineering faculty (at MIT) filed patents each year while over 60% had publications each year.

In addition to analyzing recently granted patents, entrepreneurs should track publication patterns in academic journals. Recent research (Mowery et al. 2001; Thursby and Thursby 2002) finds that patenting is usually preceded by a "flurry of publications," meaning that entrepreneurs would be well served to track the papers being published in top journals for their field of interest. In the case of patent-paper pairs, the paper will often become available before the patent application, given that most patent applications are made public 18 months after their original filing date. Thus, tracking publications in their targeted technical field will likely allow entrepreneurs to identify promising research before it is widely recognized by industry competitors, granting an advantage in developing a working relationship with the inventor.

(B) IDENTIFYING INDIVIDUAL RESEARCHERS. As the entrepreneur's attention shifts to individual researchers, it is important to understand the interplay between publishing and patenting in academia. A number of studies have examined the links between the two, and, particularly in the life sciences, found that they tend to be complements rather than substitutes. Specifically, two studies by Azoulay, Ding, and Stuart (2007 and 2009) indicate that academic patenting is linked to both higher rates of academic publication and higher publication quality. Further, analysis of the specific impact of the Bayh-Dole Act finds that while there was an increase in patenting and licensing activity at universities, there was no significant change in the content of academic research (Mowery et al. 2001; Thursby and Thursby 2002). This suggests that academics choosing to patent are doing so while pursuing projects with high academic merit, rather than instead of doing so.

Despite the existence of research which holds value in both its scientific merit and its economic utility, there remains a large fraction of senior researchers who rarely or never patent their ideas. In fact, a recent study of patenting by academics in the life sciences finds that only 11.5% of researchers are listed as inventors on one or more patents (Ding, Murray, and Stuart 2006). While it is likely that some researchers are working outside of Pasteur's Quadrant and would have little reason to patent their ideas, it is very likely that the low rate of patenting indicates significant unexplored opportunities for commercializing valuable research. This interpretation is supported by the positive relationship between patenting and publication in other studies, as well as anecdotal evidence from technology licensing offices: some scientists are simply not aware of their work's potential (Hamermesh, Lerner, and Kiron 2007).

While patenting does not necessarily impede academic research, it is important to remember that there is wide variation in researchers' willingness to engage in commercialization (particularly via entrepreneurial licensees). Research by Shane and Khurana (2003) indicates that faculty with greater experience or with higher academic rank are significantly more likely to work with newly founded firms when commercializing their inventions. Likewise, data from the life sciences suggests that female faculty are less likely to file for patents or commercialize via start-ups (Murray and Graham 2007). On the one hand, this suggests that entrepreneurs would benefit from focusing on experienced faculty when looking to gain access to university innovations, particularly when the technological field also draws considerable interest from established companies. On the other, it is also possible that junior faculty and female faculty are being systematically overlooked by entrepreneurs (and TTO professionals), who instead crowd around established faculty, leaving exciting ideas sitting on the laboratory shelf. In light of this, a strategy of specifically targeting underexplored opportunities may well increase the entrepreneur's chances of finding and securing access to a valuable invention to commercialize.

Patenting in Academia

Once a promising idea is identified, most universities ask faculty to develop a simple description of the proto-invention—an invention disclosure that covers the putative basis for the invention (i.e., what is novel, what is inventive, and why it is useful).[19] The next steps—assessment of commercial potential, patent filing, and licensing—rarely take place in a consistent and linear manner, with the precise approach varying by school. This variation arises in part because patenting is a costly and time-consuming process, and is thus highly demanding of resources: A U.S. patent can cost between $10,000 and $30,000, with even greater expenses if international IP rights are necessary. In general, TTOs do not employ internal legal experts to write patents; instead, they work with external attorneys to file patents in coordination with the inventors (who will often share the text of their academic manuscripts as the basis for the invention description—see Murray 2002).

Few TTOs file patents on all their invention disclosures. In schools with a strong culture of commercialization (e.g., Stanford, with a large research base allowing for depth of expertise across fields and long experience), TTOs (working collaboratively with faculty) typically make a rapid assessment of the likely commercial potential of an idea and file patents on a high fraction of their invention disclosures (Colyvas and Powell 2009). By contrast, at schools with a smaller research base and less-experienced faculty, or with a relatively new licensing office, assessments of patentability are often more formalized and can depend upon outside experts. At some schools, patents are only filed once a licensee is identified, running the risk that faculty are delayed in disclosing their discovery through academic channels, or that they may inadvertently disclose the discovery, thus complicating patenting efforts. It is therefore important for entrepreneurs to understand the stage of technology transfer at which a given idea resides, and whether they are engaging with a university where they need to identify ideas early in the process to ensure that ideas do not slip through the cracks unpatented.

The patenting process tends to take approximately two to five years for ideas originating in academic institutions (Hamermesh, Lerner, and Kiron 2007). Despite this delay, entrepreneurs need not wait for patents to be granted before initiating contact with the university and discussing licensing—indeed, recent research by Gans, Hsu, and Stern (2008) finds that approximately 27% of license agreements take effect before the patent is officially granted by the USPTO. This suggests that entrepreneurs must seek to identify patents, faculty, and commercial opportunities early in the transfer process. As we will describe in section 3, the process of technology transfer relies heavily on social networks and personal connections, and on the entrepreneur's ability to successfully navigate these relationships.

Accessing Intellectual Property

If we assume that an entrepreneur has identified a promising research idea at some point along the technology transfer process, he or she must now seek formal access in the form of a negotiated license. As they initiate this step, several questions are worth considering regarding access to the IP.

Question 1: Is the technology suited to an established company or an entrepreneurial one? Based on their status as small businesses, entrepreneurs do enjoy a statutory advantage in attempting to license technology under the BD Act. However, in light of the provisions and exemptions, the preference for small businesses is not an overriding one, and entrepreneurs would do well to ask licensing officers about competing corporate interests. Indeed, if there is serious corporate interest, an entrepreneur might well ask whether the technology is in fact appropriate for an entrepreneurial licensee (to form the basis of a start-up), or whether it is better suited to an established company.

To better understand how the balance between entrepreneurs and established corporations plays out in real-world licensing, we look to recent studies examining the distribution of licenses granted by top universities. The case of MIT is covered in work by Pressman et al. (1995), who find that out of a sample of 205 *exclusive* licenses, 35% were assigned to start-ups, 47% went to preexisting small businesses, and 18%

went to large firms.[20] Similarly, in a study of exclusive licenses at the University of California, Lowe and Ziedonis (2004) show that 36% were licensed to start-ups, as opposed to established (large or small) firms. In both cases, over a third of exclusive licenses flow to companies whose founding and creation is grounded in commercializing the licensed technology. In addition to capturing a considerable share of licenses, entrepreneurs have been just as successful as established firms in commercializing the licensed technology (Lowe and Ziedonis 2004). Studies also suggest that entrepreneurs and established companies generally license very different types of technology,[21] with entrepreneurs performing particularly well in younger technological fields and in industries with greater market segmentation (Shane 2001). This suggests that in areas of novel or disruptive technologies, there is significant scope for entrepreneurs to take advantage of technology transfer, even in the presence of established competitors.

Question 2: Is the IP encumbered? Even when a technology might be well suited to entrepreneurial commercialization, it is not always clear whether the IP is actually *available* to the entrepreneur. In other words, does another party already hold a right of first refusal to a negotiated license? While most IP is "free and clear" (i.e., the university has discretion to license it to whomever they deem most appropriate), some IP is already tied to another party by virtue of its initial funding. For example, if a large company has provided some research support contributing to the original invention, then as the "sponsor" of the research they would typically have the right of first refusal in license negotiations.[22] In cases when the research was subject to funding from both federal and corporate sources, critical elements of the IP are often controlled by large corporations who may be unwilling to relinquish their licensing rights. Entrepreneurs would do well to ensure that any patents they intend to license are free and clear from corporate encumbrances of this type, as such situations are becoming ever more prevalent with the rising level of corporate sponsorship across university research departments. Specific questions include whether another company has the right of refusal to a license, what the expiration period is on that right, and whether the university is willing to request a decision on that right prior to its expiry. In other instances, such as the MIT Media Lab, consortia of corporate sponsors may have the right to a nonexclusive research license, much as the government does when it funds research; this may constitute a problem for an entrepreneur seeking exclusive access to a technology.

A well-known example of a licensee with right of first refusal to license technology was DuPont's research relationship with Harvard University through funding for Professor Phillip Leder in the Department of Genetics. While his research was partly funded by grants from the NIH, DuPont's funding gave the company direct ownership of the IP over the first genetically modified animal—the Oncomouse. In this case, DuPont received a broad and exclusive patent that covered the mice themselves and methods of making them (see Murray 2010 for a longer description of the situation and the controversy that unfolded).

Question 3: Is there competing interest in the IP? In the majority of technology transfer cases, there is only one licensee that is both qualified to take advantage of the invention, and is interested in doing so (Thursby and Thursby 2004). In the rare cases where there are multiple candidates, the university may choose to create co-licensees, divide the license into multiple fields of application, or choose the single best licensee on the basis of expertise in the technology and degree of commitment to reaching the marketplace. In such scenarios, well-established connections with the university and the faculty inventor are likely to weigh heavily in the decision reached by the technology transfer office. Nonetheless, it is worth noting that competition in the context of licensing situations (particularly when seeking license exclusivity) rarely seems to involve an auction-like process in which the license is given to the highest bidder. While data on this are generally unavailable (indeed, it is not in the interest of TTOs to share aggregate levels of interest!), the market for licenses does not appear to be a "thick" market in an economic sense. Rather, as we will detail in section 3, an entrepreneur who has established a relationship with the originating laboratory (via the faculty member or junior researchers) is well positioned to work with the TTO. In addition to these relationships, however, it is often useful to sign an "option to license" with the TTO, to clearly and legally establish interest in the technology.

Licensing

Once an entrepreneur has found an idea and technology of interest, the question of licensing becomes of central importance. While TTOs vary in their strategy and structure, the use of licensing is nearly universal in the world of technology transfer, and a direct sale of a university-derived patent is extremely rare, usually occurring only in cases where the university declined to file for a patent in the first place. The details of the licensing agreement are negotiated between the TTO (acting on behalf of the university as patent owner) and the entrepreneur, often with input from the faculty member whose lab developed the technology. In cases of licensing to a start-up in which the faculty serves as founder, the faculty member can find themselves on the opposite side of the "table" from the employer. Not surprisingly faculty usually choose to allow the external entrepreneur to play a more active role in this process.

The key dimensions in this negotiation are the scope of the license, its exclusivity or nonexclusivity, and the structure of royalties and other financial obligations for the entrepreneur. Because these dimensions are fairly independent of each other, we will discuss each of them in turn.

(A) LICENSING SCOPE. The scope of the license deals with the nature of the entrepreneur's access to and use of the underlying technology behind the licensed patent. This scope can be further divided into two dimensions: technological fields and international IP rights. The first dimension applies to the (not infrequent) situations where a single patent may be the basis for a range of products in several different fields. In some cases, the licensing company may have expertise in the full range of fields covered by the invention, and may therefore wish to negotiate for the entire scope of the patent. More often, an entrepreneur licensee will only have a serious interest in a single field of application. Under such a scenario, either the university or the licensee may request that the license apply to that specific field; this would generally reduce the cost of the license, and leave the university with the option of licensing the invention's other applications to third parties. Effectively negotiating the technological scope of the license is of great importance, particularly in the case of exclusive licensing. The entrepreneur must strike a balance between keeping the scope narrow enough to control costs, and having a broad enough scope to allow for flexibility in the nature of the final product, while also protecting from competition by other licensees in related fields. This situation is well documented in Shane (2000), which describes the licensing of 3D printing technology developed in the laboratory of Professor Eli Sachs. In this case, more than half a dozen licenses were executed, each one with a different entrepreneur in an entirely distinctive field of application. Of course, it should be noted that this approach brings the entrepreneur much more limited flexibility in terms of his or her entrepreneurial strategy and chosen beachhead market.

Should s/he decide to "pivot" to alternatives, a narrow licensing scope may make this more challenging.

The geographic dimension of scope is quite similar to that of technological field. In most cases, this dimension only applies if the university has already filed for patent protection in markets outside of the U.S. In general, the entrepreneur stands to benefit from negotiating the license to cover as many countries as possible—this is particularly true if there are returns to scale in production or marketing. As in the case of technological fields, scope here has the drawback of greater costs. In this instance, the entrepreneur would likely need to cover patent application and maintenance fees across a greater range of countries. Further, the university may also negotiate responsibility for prosecuting infringement, shifting it to the entrepreneur for either international markets, domestic markets, or both. The crucial balance in this instance is between having sufficient international scope to expand to new markets if commercialization is successful, and avoiding litigation and maintenance costs in markets that are unlikely to play a large role in the long-term profitability of the product.

(B) EXCLUSIVITY.[23] The second dimension of the license negotiations is exclusivity. In virtually all cases, an exclusive license would offer greater protection from competing products, allowing for higher prices and greater profits. In light of the fact that royalties (in the biotechnology arena) average approximately 4% of net *sales* for private-sector firms licensing university inventions (Edwards, Murray, and Yu 2003), it seems reasonable to expect that any high-volume product would be more profitable if it were successfully commercialized under an exclusive license. Further, if significant additional research is required to successfully commercialize the licensed invention, an exclusive license becomes very desirable, as it offers a means of protecting the firm's research investment. Despite the important benefits above, there are several important costs associated with exclusive licenses above and beyond royalty rates. The most direct of these is that universities expect greater up-front payments and royalty rates for an exclusive license. It should be noted that universities have come under some criticism for their use of exclusive licenses, particularly for research outputs that can be considered as research tools that might benefit the entire scientific community—public as well as private. Probably the most controversial and now best-studied example derives from the exclusive license of the patents for transgenic mice to DuPont. While not an entrepreneurial start-up, the case is instructive because it highlights the challenges the licensee faced when aiming to enforce its exclusive rights over all members of the scientific community, with particular difficulties in its interactions with researchers in academia (see Murray 2010 for an extensive analysis of this case). As a result, universities typically write an "academic use" exemption into all licenses, effectively making academic institutions "out of bounds" as target customers.

When negotiating an exclusive license, entrepreneurs should also expect the inclusion of such a clause as part of the licensing agreement, specifying the commitment of funding and researchers to the commercialization process.[24] In the case of smaller firms, funding is often dependent on the ability to raise external investment capital; failure to reach the goals specified in the licensing agreement can lead the university to revoke the exclusivity of the license. In light of this, entrepreneurs should be very careful when committing to funding targets that are dependent on the results of future fundraising. In addition to contractual requirements, the greater level of university oversight may impede the ability to tightly control information about product development; this may be a cost or a benefit, depending on the nature of potential competition in the product market.[25]

(C) FINANCIAL OBLIGATIONS. Perhaps the most important dimension of a licensing agreement is the structure of its financial obligations. As expected, the total cost of the agreement is strongly dependent on the scope of the license and whether or not it is exclusive. However, the nature of payments can vary widely, and there are significant opportunities for the entrepreneur to negotiate favorable financial terms while also serving the needs of the university. This variation exists within a straightforward set of categories, each of which has different effects on risks, time horizons, and incentives between the entrepreneur and the university.

1. *Patent Costs*: The first payment category is the reimbursement of the university's patent costs. This clause is virtually standard in exclusive licenses, where universities expect full reimbursement from the licensee. In cases of nonexclusive licenses or ones of limited scope of application, the entrepreneur should expect to only partially reimburse these costs.
2. *Up-Front Fees*: The second category is the up-front fee that the university demands for issuing the license. This fee is meant to be a source of direct rev-

enue for the university and its researchers. Entrepreneurs should be wary of committing significant up-front payments to inventions that require significant time, risk, and investment before commercial viability.

3. *Maintenance Fees and Milestone Payments*: In cases where significant amounts of research and product development remain, the entrepreneur should focus on the third payment category: license maintenance fees and milestone payments. These have the effect of spreading payments over time and importantly, this financial structure better aligns incentives for the entrepreneur and the university to work together.
4. *Royalties*: The fourth category covers conditional payments—royalties on the products sold by the licensee. Royalties are almost always calculated as a percentage of sales of final goods or services. They depend heavily on the stage of development of the invention, the magnitude of the needed contribution from the entrepreneur, the profit margin expected for the final product, and on whether other licenses are required to create a viable product. In biotech, average royalty rates are approximately 4% of sales (Edwards, Murray, and Yu 2003).
5. *Equity*: Perhaps the most important and perplexing aspect of the licensing agreement for any novice entrepreneur working with a university lies in the role of equity. At its core, equity serves as a deferred payment (in lieu of large up-front fees), whose value is conditional on successfully commercializing the invention. Equity shares therefore provide a strong incentive for the university to offer support to the entrepreneur; further, using equity as a method of payment can leave the entrepreneur's financial assets free for investment into the research and equipment necessary for commercialization. What is important to explore for the entrepreneur is the typical rate of equity participation at the university in question. While such statistics are rarely found in published data, it is generally understood that universities vary widely in their equity requirements. Some of the oldest TTOs in the U.S. take low single-digit equity rates, while others more intent on participating in liquidity events take a larger share. Those universities with more active engagement in company formation, which seek to offer more than transactional IP licensing, will also often take more equity. In some instances, particularly for academic medical centers, accepting equity is problematic, as it could lead to conflicts of interest; these would arise when and if patients were to use products commercialized by companies in whom the hospital has equity. Further, the university may find equity to carry more risk than royalties, and because revenues tend to rise before a company becomes profitable, the university will likely be forced to wait longer before benefiting from this form of payment.

In this section, we have described the details of the technology transfer process at the level of the university. The discussion has focused primarily on the role of the technology transfer office, and on the details of the licensing agreement. The dimensions and tradeoffs that make up any given licensing agreement are reasonably uniform; however, it is important to remember that the details of a standard contract can vary widely, both between different technological fields and between universities. These details are often proprietary information, meaning that measures such as royalty rates or the division between up-front fees and maintenance payments are not publicly available. Once the entrepreneur has focused in on a specific technology and university, the best source of information for the expected level of payments is usually recent licensees who have negotiated over comparable technologies with the same licensing office. In the case of new technological fields, or universities which are just starting to contribute to an existing area of research, there may not be any licensees who match both the school and the technology. Despite this, the above approach continues to be valuable: the entrepreneur can help set expectations by contacting recent licensees who have worked with the school in a different field, or worked with similar technologies elsewhere. However, in such cases, the entrepreneur should expect to devote a reasonable amount of time to adjustment and education as the school's licensing office performs its due diligence and gains experience during the first stages of negotiations.

3. Individual Incentives and Personal Networks

In the previous sections, we first described the legislative framework governing technology transfer under the Bayh-Dole Act in the United States, and then discussed the general process of working with a university TTO to identify, access, and license promising IP. In this section, we focus in on the role of personal interactions in ensuring effective commercialization; we focus primarily on interactions between the lab and the entrepreneur, and highlight the importance of individual motivations and the role of social networks. Building on our discussion of licensing in the previous section, we begin with the formal, license-based incentives shaping the engagement of faculty (and other research lab members) in identifying promising research and pursuing technology transfer. We then discuss the importance of informal and nonmonetary incentives in the early stages of engagement with faculty, before moving chronologically to the incentives and individual roles that pertain during commercialization and product development. Finally, we move past incentives and organizational structures to focus in on the finest level of detail: social networks and personal relationships. This allows us to close by emphasizing the most focused and likely the most important aspects of the entrepreneur's role in the commercialization process.

Incentives for Technology Transfer

LICENSE-BASED INCENTIVES: While the path of scientific ideas from invention disclosure to patent to licensing and venture formation may seem straightforward and entirely disjoint from the daily life of the laboratory, in fact, the participation of faculty and other lab members in the transformation is critical. Useful scientific ideas are simply that—ideas—and without the wealth of tactic knowledge and experience they can be difficult to move from the confines of the academic laboratory bench into the real world. Indeed, recent research suggests that providing this soft information is just as significant as publishing patents when analyzing the economic and innovative impact of academic researchers (Agrawal and Henderson 2002). Both the university and the entrepreneur must offer strong motivation to encourage researchers to devote their time and effort to the technology transfer process. At first glance, this seems to be a significant obstacle: researchers are under no obligation to participate in technology transfer. Further, the central goal of receiving tenure is based almost exclusively on publications and grants rather than the researcher's history of patenting and commercialization (Siegel, Waldman, and Link 2003). In addition, researchers face a high opportunity cost of their time—both their career trajectory and their standing within their discipline is driven by the quantity and quality of their publications, which, in addition to the demands of teaching and mentoring, leave very little remaining time for secondary interests.

To offer an incentive for technology transfer and commercialization, the Bayh-Dole Act specifies that a portion of licensing revenues must go directly to the original inventor(s). Because of this, the university and its TTO rely on the licensing agreement as the fulcrum for structuring incentives. In many cases, licensing revenues are divided into three equal parts, respectively flowing to the university, the academic department overseeing the research, and the original inventor(s).[26] As the source of licensing revenue, the entrepreneur has significant control over the timing and structure of the license-based incentives available to faculty researchers. During license negotiations, the entrepreneur should consider structuring a contract which favors maintenance fees and milestone payments over up-front fees, as it provides a consistent source of incentives for the original inventor to collaborate in commercialization. While royalties will also lead to a one-third share of payments flowing to the original researcher, these payments usually accrue only after commercialization is complete, and may seem too distant or uncertain to provide strong incentives during the early stages of technology transfer. With careful planning, the entrepreneur can negotiate a payment structure that provides continuous formal incentives for faculty cooperation.

BROADER MOTIVATIONS FOR LAB ENGAGEMENT: In general, the inventor's share of licensing revenues will tend to be comparable in magnitude (in cases of successful commercialization) to alternative sources of income such as grants and academic salaries. In most cases, this will not be large enough to compensate the researchers for more than a fraction of his or her actual working hours engaged in commercialization.[27] Thus, entrepreneurs must offer researchers a range of valuable opportunities that are difficult to pursue within a purely academic environment. In this section, we highlight the two categories of incentives that are most likely to be influential when working with academics: real-world impact and prestige.

The first significant source of incentives comes from researchers' desire to have a meaningful impact with their inventions: while their academic work may be very influential within their field, many researchers place significant value on having a broader impact. While many researchers recognize and worry about potential conflicts between commercial activity and academic research, these are often seen as a challenge one must overcome in order to make a difference in the world. As one researcher explained, "I raised it with my mentors, who assured me it was a conflict I had to manage as [it] was essential to further my science and really make an impact" (Murray and Graham 2007). Because of this desire on the part of researchers, entrepreneurs can offer significant value by developing the researcher's invention into a commercial product, where it can reach a broader population and often satisfy previously unaddressed needs of consumers or other beneficiaries. For many researchers, this will align closely with their desire for making a real-world impact, leading to the buy-in that is central to successful commercialization.

The second source of incentives is rooted in one of the primary currencies of the academic world: prestige and recognition. For a researcher, reputation is an important determinant of success during the tenure process, but in relative terms, it becomes an even more central source of motivation once tenure has been granted. The desire for reputation often focuses on concrete measures, including attributions such as coauthorship or citations to their publications (Gans and Murray 2013). Along these lines, patenting and commercializing an invention offers a source of recognition through follow-on patents, which are likely to cite both the patent describing the original invention and the academic publications which led to it. Further, commercialization can raise the visibility of the underlying research, leading to more academic citations, particularly if the scope of applicability expands during product development. Both the citations and the direct visibility gained through technology transfer lead to an improvement in professional reputation for the original inventor and the collaborators on the project. In addition to this, commercialization offers recognition in a general context, beyond the researcher's narrow technological field. This last attribute is particularly advantageous for the entrepreneur, as this form of recognition is very difficult to achieve in a purely academic setting, and allows the researcher to more easily explain their work to broader audiences. By supporting researchers' desire for recognition both within and outside their technological field, the entrepreneur provides a significant source of value and incentives, fostering the collaboration necessary for successful commercialization.

Incentives for Commercialization

So far, we have described some of the incentives for faculty and student participation in the narrow process of technology transfer. However, as noted above, commercialization requires much deeper engagement and interaction between researchers with significant technical expertise and reputation, and the entrepreneur with business expertise and access to external capital and ongoing resources. The incentives for this deeper engagement are driven by both financial motives and a desire for impact, and should be structured around the specific role that each lab member plays in the commercialization process.

One of the first steps the entrepreneur should take is to explicitly define the role of the original faculty-inventor in the start-up (see Murray 2004 for a more detailed discussion of the role of faculty in entrepreneurial start-ups). Importantly, the inventor is rarely the CEO: despite his or her expertise at running a research lab, this expertise is unlikely to translate well to other important aspects of a start-up.[28] Specifically, tasks such as working with investors, organizing production and marketing, and engaging with customers to calibrate the final product to their needs are unlikely to be a good fit for the inventor, and are best handled by the entrepreneur or the corresponding experts at the start-up. Moreover, many schools have prohibitions on faculty taking a formal executive role unless on sabbatical or on a leave of absence. Rather than taking a managerial position, the most fitting role for the inventor can be that of an advisor, either as a member of a scientific advisory board (SAB) or, in cases of previous experience, as a member of the board of directors for the start-up. Occasionally faculty will supplement this role by serving (again for a limited period) as the chief technology officer during a leave or sabbatical. In most cases, the entrepreneur's goal is to avoid burdening the inventor with the daily details of the company; this leaves them free to focus on the underlying technology and ways in which it could be improved, while also allowing them to focus on the next academic project. At the same time, this allows the entrepreneur to handle aspects related to management and strategic vision; this division of responsibilities plays to the strengths of both parties, leading to both personal fulfillment and a greater chance of successful commercialization.

In addition to defining the role of the faculty member, the entrepreneur should seek out a team of junior researchers (potentially including the graduate student who led the research) who can ultimately devote a significant portion of their time to the details of commercialization; ideally, they would join the venture to lead the transfer process and the venture's technology development. Importantly, this team should already be familiar with the technology in question; therefore, they are best drawn from the lab and department of the original inventor. Because of this, the entrepreneur typically defers to the inventor's advice in putting together this research team, in order to benefit from the inventor's first-hand knowledge of his junior researchers and to minimize any friction resulting from conflicts between commercial interests and academic research. Ideally, at least some of the graduate students and junior researchers will transition to full-time positions at the start-up, in order to focus all of their attention on commercialization.

The entrepreneur can and typically does offer significant financial incentives for the researchers' expertise and assistance. These incentives are highly contingent upon the details of the role that lab members—the faculty, students, and postdocs—play in ongoing commercialization. As a general rule, faculty who play any sort of role in conceptualizing and commercializing a technology from their laboratory will serve as founders in a new venture and thus take some share in the founder's equity.[29] Despite a history of criticism in some areas of academia, the strategy of granting equity to faculty researchers has become commonplace in technology transfer. A recent study found that in 2004, 50% of IPO-stage companies had granted equity to faculty, accounting for over 8% of the IPO valuation (Edwards, Murray, and Yu 2006). Because the value of equity grows with the success of commercialization it offers excellent incentives for researchers, albeit at a significant cost to the entrepreneur's equity share. Because of this tradeoff, the level of faculty founder's equity should reflect the importance of the original inventor's initial and ongoing expertise during commercialization. In addition to the primary faculty member (who likely remains "behind" in the university), founder's equity can also be granted to the central technical individual from the lab (or elsewhere) who is driving commercialization, particularly if they leave the university to do so (see Wasserman 2012 for a discussion of equity splits).

Personal Networks

Having outlined the set of incentives that the university and the entrepreneur can put in place to drive commercialization, the primary remaining question relates to how an entrepreneur can position him or herself to be *the* individual poised to build a strong foundation for collaboration with the original inventor. How can they ensure they are the one driving commercialization, and not another entrepreneur? In simple terms this relies upon social networks, expertise, and experience and is best understood along the path of development from idea identification to licensing.

Given the uneven distribution of useful ideas across disciplines, departments, and labs, how are potential opportunities actually identified by entrepreneurs? The most obvious way might seem to be via Google and the U.S. patent office![30] However, in most schools, particularly in today's climate of attention on commercialization, entrepreneurs usually seek to build (and maintain) close relationships with faculty researchers and with TT officers.[31] This allows them to identify promising research at a very early stage and ensure that invention disclosures and patent applications are filed in a timely manner, so as not to lose intellectual property rights via widespread disclosures (at conferences, academic symposia, or via early publication).

In general, the earlier the entrepreneur engages with the idea or invention, the more effective the relationship that will develop. Having early knowledge of emerging technologies is an important asset for entrepreneurs, allowing them to initiate contact and explore commercialization opportunities well in advance of the patent grant date. Several paths are open to the entrepreneur and should probably be followed simultaneously. First, TTOs are generally available for meetings with potential licensees, and the licensing officers responsible for particular departments or faculty will have extensive knowledge of particular technology domains and the active research within them. In many cases, universities will also actively seek licensees for inventions immediately after submitting a patent application, often through a list of available technologies on their website.[32] Second, whether or not the technology transfer office is proactive, it is still possible (and important) to obtain information on the technology: all patent applications become publicly available through the USPTO 18 months after the initial filing date. The most important goal for the entrepreneur is to identify particular faculty members and research groups working in the entrepreneur's field of interest. This identification sets the stage for the entrepreneur to focus in on a specific research domain, and seek to gain a deeper understanding of the activities of the faculty, including specific projects, areas of active research, publications, and existing and forthcoming patents. In turn, this is likely to form the foundation for productive relationships with the original inventor, and can provide significant benefits when interacting with the TTO.

Once the entrepreneur has identified a valuable invention or a researcher whose work merits further exploration, the most important step is developing an effective relationship with the inventor. Doing so without any shared connections or prior contact is unlikely to be successful. Instead, the entrepreneur is best served by building up a set of bridge contacts, with the goal of eventually reaching the head researcher of interest. These contacts can take many forms: they can be other senior academics with whom the entrepreneur has an established relationship, or graduate students, junior researchers, or even other entrepreneurs who are connected to the head researcher. Personal connections can be some of the most valuable assets in the worlds of venture capital and entrepreneurship, and technology transfer is no exception. In both biotechnology and venture capital, personal ties and geographic proximity are important determinants of overall success (Lindqvist 1991; Sorenson and Stuart 2001; Powell et al. 2002). In the previous sections, we have noted the value of these connections: close ties with a licensing office can help an entrepreneur in during both the search for promising inventions and the licensing process, and connecting with other entrepreneurs can help set expectations for negotiating a contract with the university. However, the most important connection in technology transfer is that between the entrepreneur and the academic researcher.

A strong social relationship between the entrepreneur and the lab can be valuable throughout the commercialization process. After identifying ideas and prompting effective invention disclosure, social relationships can ensure that the entrepreneur is best positioned to license the invention: in many cases, a school's TTO will (informally) grant preference to the inventor's "commercialization team" if there are multiple parties interested in a given patent. At other schools, the office will take a stronger role in determining the recipient of the licensing contract; in these cases, contacts with the licensing officers and connections to previous licensees will help establish a foundation for effective negotiations. In general, while the BD Act's regulatory preference for small businesses has the potential to benefit the entrepreneur, the realization of this benefit depends largely on establishing a strong network of contacts at the university's technology transfer office, and with the faculty researchers and labs whose invention the entrepreneur seeks.

Despite its central role in technology transfer, signing a licensing agreement is only the first step toward commercialization; in most cases, there are significant challenges in taking an academic invention and developing a marketable product. The ability to overcome these challenges depends crucially on being able to take advantage of a broader set of social networks. Indeed, recent research indicates that conditional on obtaining a license to an emerging technology, one of the strongest determinants of the ultimate success of the venture is the social capital of the founder (Shane and Stuart 2002). Thus, in addition to formal contracts and incentives, networks and the personal trust they build can help to develop a productive relationship with the inventor and, crucially, help to ensure faculty engagement during the commercialization process. Fortunately, the range of contacts developed over the identification and licensing stages of technology transfer are likely to offer an effective foundation for facilitating buy-in with the original inventor and his team of researchers. By building strong relationships with researchers, entrepreneurs will have access to the full range of knowledge and expertise needed for successful commercialization. In addition to fostering these relationships, the entrepreneur should continually be on the lookout for new connections: while the beginning of the technology transfer process depends on effective use of social networks in the specific academic field and university of the invention, over time these networks will need to expand to include a range of other entrepreneurial team members, early-stage investors, advisors, those with access to other critical resources.

Conclusions

In this chapter, we have discussed the details of technology transfer from the perspective of the entrepreneur. We began with a discussion of national institutions, focusing on the Bayh-Dole Act and the legal aspects of technology transfer in the United States. In doing so, we highlighted the responsibilities of the entrepreneur and the pitfalls to avoid when seeking to license academic inventions. Next, we turned to local norms and interpretations at the level of individual universities, focusing on the role of technology licensing offices. We described the variation in licensing strategies across universities, and then turned to the details of the licensing contract and the negotiating process between the entrepreneur and the university. Finally, we examined the role of networks and individuals in effective commercialization, describing the nature of incentives and institutional roles through which the entrepreneur can most effectively collaborate with the original inventor, and then emphasizing the importance of personal connections throughout the technology transfer process. In doing so, we have created a framework for the entrepreneur's decisions over the course of technology transfer, explaining the tradeoffs at each stage and offering general guidance while allowing the entrepreneur to remain flexible and responsive to conditions on the ground throughout the commercialization process.

Notes

1. For more details on the Eta Devices story, see http://etadevices.com/press/release/.

2. For information of the detailed requirements of the Bayh-Dole Act, see the information contained in http://www.ucop.edu/ott/faculty/bayh.html. Similarly, the details of the Stevenson-Wydler Act are available at http://www.csrees.usda.gov/about/offices/legis/techtran.html.

3. Data from the NSF's 2011 Higher Education Research and Development Survey, available at http://www.nsf.gov/statistics/srvyherd/.

4. From http://www.law.cornell.edu/uscode/text/35/200.

5. Not all universities file patents under the university name, e.g., the University of Wisconsin files as the "Wisconsin Alumni Research Foundation" for historical reasons. Foreign universities also use different naming conventions; for example, the University of Oxford files via its subsidiary Isis Innovation.

6. http://www.biotherapharma.com/press/2007/7-27-07-Biorgin.html.

7. The patent was reissued in 1999 as U.S. Patent No. RE 36,478, "Processing of acoustic waveforms."

8. http://tech.mit.edu/V123/N20/2010ckheed.20n.html.

9. http://www.law.cornell.edu/uscode/text/35/203.

10. Such concerns were determined to be better addressed either by an investigation of potential anticompetitive practices through the FTC, or by legislative acts of Congress.

11. http://www.ucop.edu/ott/faculty/bayh.html#Current Regulations.

12. http://web.mit.edu/tlo/www/about/our_mission.html (accessed 9/9/2012).

13. The WARF was founded in 1925 to commercialize discoveries in the synthesis of Vitamin D—essential to curing rickets—made by Wisconsin professor Harry Steenbock.

14. http://www.warf.org/about/index.jsp?cid=26 (accessed 9/9/2012).

15. Note that for *international* IP rights, the patent application must be filed *prior* to publication or disclosure of any kind.

16. It is unreasonable to expect perfect foresight in identifying promising research; this section therefore emphasizes evaluating potential, rather than realized outcomes.

17. His analysis of the germ theory of disease not only expanded the scientific understanding of microbiology, but also led to the development of vaccines for rabies and anthrax, and the process of pasteurization to prevent spoilage in milk and similar products.

18. Even if a single project yields patentable ideas that are not ultimately patented, the large ratio underscores the average scarcity of dual research in many universities. (See Gans, Murray and Stern 2011 for an equilibrium analysis of the disclosure strategy followed by researchers and their funders.)

19. Recent research indicates that not all inventions are commercialized through TTOs. Instead, some move out of the university and into the private sector through "backdoor" paths, by direct interaction between researchers and industry (Link et al. 2007).

20. Interestingly, while start-ups represent only 35% of exclusive licenses at MIT, they account for 77% of total investment induced by licensing (Pressman et al. 1995).

21. The main exception to this involves cases where both entrepreneurs and established firms pursue a nonexclusive license to a platform technology with widespread applications.

22. See Gans and Murray (2011) for more detail.

23. This section draws heavily from Hamermesh, Lerner, and Kiron (2007), appendices B and C.

24. Note that while less common, diligence requirements can also be found in non-exclusive licensing contracts.

25. While this topic has not been studied extensively, one might expect unplanned disclosures of information to encourage potential competitors early in the commercialization process. By contrast, when commercialization is nearly complete, disclosure of this progress may well serve to deter competition.

26. An example of the explicit division of license proceeds can be seen at http://www.albany.edu/osp/24629.php.

27 There are occasional exceptions to this trend, such as the research by Brian Seed on the pharmaceutical drug Enbrel, developed at Massachusetts General Hospital and co-marketed by Amgen and Pfizer.

28. In specific instances either when the faculty has no mechanism of seeking out an entrepreneur they trust or when they are less familiar with entrepreneurial teams, a faculty member will serve as a CEO for a short period.

29. From the previous section, we know that the TTO will typically take on an equity share in the company as part of licensing negotiations, with a portion of this revenue flowing back to the inventor. While this appears to be a form of double-dipping, it is important to remember that any equity going to the faculty or lab member in recognition of their role as a founder is a transaction entirely independent of the licensing office, and one that would take place regardless of the technological origins of the company.

30. U.S. patent applications at least 18 months old and granted U.S. patents can be searched at the following website: http://www.uspto.gov/patents/process/search/index.jsp.

31. The presence of strong relationships between industry and technology licensing offices is a primary determinant of the ability to identify and develop higher-impact patent portfolios (Owen-Smith and Powell 2003).

32. An example of such a list of technologies is the online directory maintained by Stanford's Office of Technology Licensing, available at http://techfinder.stanford.edu.

References

Agrawal, A., and R. Henderson. 2002. "Putting Patents in Context: Exploring Knowledge Transfer from MIT." *Management Science* 48 (1): 44–60.

Azoulay, P., W. Ding, and T. Stuart. 2007. "The Determinants of Faculty Patenting Behavior: Demographics or Opportunities?" *Journal of Economic Behavior and Organization* 63 (4): 599–623.

Azoulay, P., W. Ding, and T. Stuart. 2009. "The Impact of Academic Patenting on the Rate, Quality, and Direction of (Public) Research Output." *Journal of Industrial Economics* 57 (4): 637–76.

Cohen, S., et al. 1973. "Construction of biologically functional bacterial plasmids in vitro." *PNAS* 70 (11): 3240–44.

Colyvas, J., and W. Powell. 2009. "Measures, metrics, and myopia: The challenges and ramifications of sustaining academic entrepreneurship." *Advances in the Study of Entrepreneurship, Innovation and Economic Growth* 19: 79–111.

Ding, W., F. Murray, and T. Stuart. 2006. "Gender Differences in Patenting in the Academic Life Sciences." *Science* 313 (5787): 665–67.

Duecker, K. 1997. "Biobusiness on campus: Commercialization of university-developed biomedical technologies." *Food and Drug Law Journal* 52: 453–510.

Edwards, M., F. Murray, and R. Yu. 2003. "Value Creation and Sharing Among Universities, Biotechnology, and Pharma." *Nature Biotechnology* 21 (6): 618–24.

Edwards, M., F. Murray, and R. Yu. 2006. "Gold in the Ivory Tower: Equity Rewards of Outlicensing." *Nature Biotechnology* 24 (5): 509–15.

Gans, J., D. Hsu, and S. Stern. 2008. "The Impact of Uncertain Intellectual Property Rights on the Market for Ideas: Evidence from Patent Grant Delays." *Management Science* 54 (5): 982–97.

Gans, J., and F. Murray. 2013. "Markets for Scientific Attribution." Mimeo, MIT.

Gans, J., F. Murray, and S. Stern. 2010. "Contracting Over the Disclosure of Scientific Knowledge: Intellectual Property and Academic Publication." SSRN Working Paper. Available at http://papers.ssrn.com/s013/papers.cfm?abstract_id=1559871.

Guo, T., C. Jin, and R. Smalley. 1991. "Doping bucky: Formation and properties of boron-doped buckminsterfullerene." *Journal of Physical Chemistry* 95 (13): 4948–50.

Ham, R., and D. Mowery. 1998. "Improving the effectiveness of public-private R&D collaboration: Case studies at a US weapons laboratory." *Research Policy* 26(6): 661–75.

Hamermesh, R., J. Lerner, and D. Kiron. 2007. "Technology Transfer at U.S. Universities." Harvard Business School Publishing: HBS Case 9-807-124.

Lindqvist, M., 1991. *Infant Multinationals: The Internationalization of Young, Technology-Based Swedish Firms.* Stockholm School of Economics, doctoral thesis.

Link, A. N., D. S. Siegel, and B. Bozeman. 2007. "An Empirical Analysis of the Propensity of Academics to Engage in Informal University Technology Transfer." *Industrial and Corporate Change* 16 (4): 641–55.

Lowe, R., and A. Ziedonis. 2004. "Start-Ups, Established Firms, and the Commercialization of University Inventions." Working Paper.

Mowery, D. C., et al. 2001. "The Growth of Patenting and Licensing by U.S. Universities: An Assessment of the Effects of the Bayh-Dole Act of 1980." *Research Policy* 30: 99–119.

Murray, F. 2002. "Innovation as Co-evolution of Scientific and Technological Networks: Exploring Tissue Engineering." *Research Policy* 31 (8–9): 1389–1403.

Murray, F. 2004. "The Role of Academic Inventors in Entrepreneurial Firms: Sharing the Laboratory Life." *Research Policy* 33: 643–59.

Murray, F. 2010. "The Oncomouse That Roared: Hybrid Exchange Strategies as a Source of Distinction at the Boundary of Overlapping Institutions." *American Journal of Sociology* 116 (2): 341–88.

Murray, F., and L. Graham. 2007. "Buying Science and Selling Science: Gender Differences in the Market for Commercial Science." *Industrial and Corporate Change* 16 (4): 657–89.

Murray, F., and S. Stern. 2007. "Do formal intellectual property rights hinder the free flow of scientific knowledge? An empirical test of the anti-commons hypothesis." *Journal of Economic Behavior and Organization* 63 (4): 648–87.

Owen-Smith, J., and W. W. Powell. 2003. "The expanding role of university patenting in the life sciences: Assessing the importance of experience and connectivity." *Research Policy* 32: 1695–1711.

Powell, W. W., et al. 2002. "The Spatial Clustering of Science and Capital: Accounting for Biotech Firm: Venture Capital Relationships." *Regional Studies* 36 (3): 291–305.

Pressman, L., et al. 1995. "Pre-production investment and jobs induced by MIT exclusive patent licenses: A preliminary model to measure the economic impact of university licensing." *Journal of the Association of University Technology Managers* 7: 49–82.

Sachs, E., et al. 1990. "Three-Dimensional Printing: Rapid Tooling and Prototypes Directly from a CAD Model." *CIRP Annals—Manufacturing Technology* 39 (1): 201–4.

Shane, S. 2000. "Prior Knowledge and the Discovery of Entrepreneurial Opportunities." *Organization Science* 11 (4): 448–69.

Shane, S. 2001. "Technology regimes and new firm formation." *Management Science* 47 (9): 1173–90.

Shane, S., and R. Khurana. 2003. "Bringing individuals back in: The effects of career experience on new firm founding." *Industrial and Corporate Change* 12 (3): 519–43.

Shane, S., and T. Stuart. 2002. "Organizational Endowments and the Performance of University Start-ups." *Management Science* 48 (1): 154–70.

Siegel, D., D. Waldman, and A. Link. 2003. "Assessing the Impact of Organizational Practices on the Relative Productivity of University Technology Transfer Offices: An Exploratory Study." *Research Policy* 32 (1): 27–48.

Siegel, D., and M. Wright. 2015. "University Technology Transfer Offices, Licensing, and Start-Ups." In *Chicago Handbook of University Technology Transfer*. Chicago: University of Chicago Press.

Sorenson, O., and T. Stuart. 2001. "Syndication Networks and the Spatial Distribution of Venture Capital Investments." *American Journal of Sociology* 106 (6): 1546–1588.

Stokes, D. 1997. *Pasteur's Quadrant: Basic Science and Technological Innovation*. Washington, D.C.: Brookings Institution Press.

Thursby, J., and M. Thursby. 2002. "Who is Selling the Ivory Tower? Sources of Growth in University Licensing." *Management Science* 48: 90–104.

Thursby, J., and M. Thursby. 2004. "Are Faculty Critical? Their Role in University-Industry Licensing." *Contemporary Economic Policy* 22 (2): 162–78.

Wasserman, N. 2012. *The Founder's Dilemmas: Anticipating and Avoiding the Pitfalls that Can Sink a Startup*. Princeton, NJ: Princeton University Press.

CHAPTER FIVE

Challenges in University Technology Transfer and the Promising Role of Entrepreneurship Education

Andrew Nelson and Thomas Byers

Introduction

The purpose of this chapter is to reflect upon the challenges that confront university technology transfer efforts and to consider the role of entrepreneurship education both in addressing these challenges, and as an area in its own right. To begin, we outline the boundary conditions for our reflections, noting the somewhat narrow domain of university technology transfer through start-ups. We then expand upon the general challenges for commercialization of new technologies, focusing especially on resource requirements that are exaggerated in the case of university spin-outs. Finally, we consider the role of entrepreneurship education in addressing these challenges. We propose a number of specific ways in which entrepreneurship education can proceed most effectively while suggesting several important distinctions between entrepreneurship education and technology transfer that, we argue, are important to recognize and maintain.

The Domains and Intersection of University Technology Transfer and Entrepreneurship Education

Any investigation of university technology transfer and entrepreneurship must begin with an acknowledgement that most university technology transfer does not proceed through start-ups and spin-outs. As illustrated in figure 5.1, annual surveys conducted by the Association of University Technology Managers (AUTM) indicate that established companies, not start-ups, hold the vast majority of licenses to university technologies—a fact that has remained relatively consistent over time. A technology-licensing officer with a major U.S. university shared with us one explanation for the high proportion of established companies among all licensees:

> When I review a new [invention] disclosure and am attempting to figure out if there's a market for it and if we can find a licensee, the natural place to look is the set of companies that have licensed similar things from us in the past. Almost by definition, that means that we're looking first to more established companies, unless the inventor specifically points us to a startup. But even then, we're still going to check with our existing portfolio [companies] and contacts to assess interest, figure out the market value, and so on.

In light of this situation, the heavy emphasis on start-ups, in connection to university technology transfer, may come as a surprise (e.g., Degroof and Roberts 2004; Di Gregorio and Shane 2003; Franklin, Wright, and Lockett 2001; Lerner 2005; Lockett, Wright, and Franklin 2003; Lockett and Wright 2005; Markman et al. 2005; Nerkar and Shane 2003; O'Shea et al. 2005; Siegel, Wright, and Lockett 2007; Smilor, Gibson and Dietrich 1990; Wright, Birley, and Mosley 2004).

Start-ups and large companies cannot be considered simple substitutes for one another in the commercialization of university research, however, for four reasons. First, Tushman and Anderson (1986) and Henderson and Clark (1990) outlined how disruptive technologies and "architectural" innovations require different capabilities and strategies to commercialize than do incremental technologies. In turn, Christensen (1997) proposed that established companies face an "innovator's dilemma" in that they do not wish to erode existing market share by promoting new and disruptive technologies. As a result, they are less likely than start-ups to attempt to commercialize these types of technologies. Insofar as universities are more likely to conduct basic research that leads to disruptive rather than incremental technologies, start-ups may play a crucial and disproportionately active role in commercializing important disruptive technologies. In turn, this role may not be borne out in statistics on volume alone.

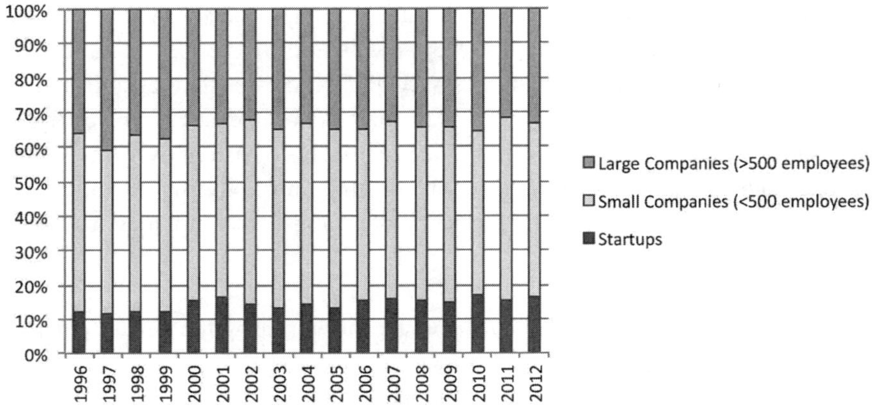

FIGURE 5.1. Percentage of university technology licenses signed with start-ups, small companies, and large companies over time. Source: AUTM 2013.

Second, many universities, especially those that are public, emphasize local or regional impact as part of their mission. Insofar as start-ups are locally based and large companies are not, an emphasis on university technology transfer through start-ups is understandable (Benneworth and Charles 2005; Feldman and Desrochers 2003; Siegel and Phan 2005).

Third, on a related note, it is easier to assess economic impact through start-ups versus large companies. For example, it is straightforward to claim that a license facilitated a new start-up that now employs eight people. It may also be the case that the same license would have permitted a large company located in the same region to add eight employees (or to save eight employees from losing their positions). Absent a detailed understanding of the internal labor market and organization within this large company, however, the case for regional employment and economic impact is more difficult to establish in the large company case.

Finally, as one university administration official shared with us, "Startups are in; big corporations are not. We'd much rather tell community leaders, donors, alumni and others, 'We support startups,' than, 'We support multinationals.'" Thus, positive social affect toward start-ups may explain, in part, the emphasis on start-ups. For all of these reasons, and undoubtedly many more, discussion of start-ups dominates much of the academic literature on university technology transfer, even though start-ups represent a fraction of university technology transfer activities.

Another aspect of our investigation of university technology transfer and entrepreneurship education is that we direct our comments specifically at the "traditional" case of start-ups who sign licenses for university inventions. In so doing, we fully recognize that these formal arrangements fail to capture a great deal of other links between technology transfer and entrepreneurship. For example, Agrawal and Henderson (2002) surveyed faculty in MIT's departments of mechanical and electrical engineering. Their results indicated that patents and licenses account for only 7% of knowledge flows out of these researchers' labs. In an exhaustive study of Stanford's computer music center, which ranks among the most active out-licensors and patentors at Stanford, Nelson (2012) found that licenses captured only about 5% of the organizations developing these technologies for commercial use (as indicated by patent citations). Both Agrawal and Henderson (2002), as well as Nelson (2012), show that far more technology transfer takes place via conference presentations, publications, hiring of recent graduates, consulting, and collaborative research than via development by license-holding university spin-outs.

Finally, as we discuss further in the final section of this chapter, entrepreneurship, in our view, extends far beyond "starting a new company." Following Shane and Venkataraman (2000: 218), we argue that entrepreneurship fundamentally is concerned with "how, by whom, and with what effects opportunities to create future goods and services are discovered, evaluated, and exploited." Thus, entrepreneurship may involve social rather than (or in addition to) commercial aims (e.g., Mair and Marti 2006; Martin and Osberg 2007; Tracey and Phillips 2007) and it may take place within existing organizations alongside start-ups (e.g., Antoncic and Hisrich 2001; Antoncic and Hisrich 2003; Carrier 1993; Pinchot 1985). Clearly, this conceptualization extends the domain of entrepreneurship education beyond the training of students on "how to start a new company" and it signals that university technology transfer efforts represent only a small segment of entrepreneurship education.

Taken together, these parameters indicate that our exploration of the intersection of university technology transfer and entrepreneurship education must be acknowledged as capturing only a small fraction of each of these activities. Nevertheless, given the important role of university spin-outs in innovation and economic growth (Acs, Audretsch, and Feldman 1994; Audretsch, Keilbach, and Lehmann 2006; Audretsch and Stephan 1996; Bania, Eberts, and Fogarty 1993; Hart 2003; Mueller 2006; Wennekers and Thurik 1999) alongside the many difficult challenges with these spin-outs, it is critical to consider the ways in which entrepreneurship education might enhance the effectiveness of these efforts.

A Review of the Resource Challenges for University Spin-Outs

As a number of scholars have noted, commercialization of research presents both technology risks and market risks (e.g., Byers, Dorf, and Nelson 2014; Kaplan and Strömberg 2004; Rosenberg 1996; Shepherd, Douglas, and Shanley 2000). Technology risks center on the fundamental question, "Can you build it?" The answer to this question must address issues of reliability, ease of mass production, cost to manufacture, and so on. Market risks, parroting the famous line from the movie *Field of Dreams*, center on the question, "If you build it, will they come?" A strong answer to this question must address whether the market will develop quickly enough, in large enough numbers, with customers willing to pay the necessary sales price, and so on. Both technology and market risks loom very large in the case of research commercialization. Arguably, however, these challenges are not unique to the university case or to the nascent start-up case; any company that attempts to commercialize a new technology faces these same risks. We focus our comments in this section, therefore, on the unique resource requirements that a university spin-out confronts.

The most conspicuous resource in the minds of many entrepreneurs is financial. As the adage goes, "It takes money to make money." In the case of start-ups, financial resources may be necessary to develop a product, attract a team, conduct marketing activities, secure a facility, and perform other basic functions associated with building a company. In this context, practitioners and academic researchers alike focus considerable attention on venture capital as a finance mechanism (Baum and Silverman 2004; Hallen 2008; Jeng and Wells 2000; Kortum and Lerner 2000; Wright, Vohora, and Lockett 2004; Wright et al. 2006). It is worth noting, of course, that only a small percentage of start-ups exhibit the scalability and large market potential to make them good candidates for venture capital and that an even smaller percentage of start-ups ever receive venture capital. For example, Kauffman Firm Survey data indicate that among 3,564 capital injections received by firms in the 2004 survey, just 26 were in the form of venture capital (less than 1%). Of those firms that survived to 2008, just seven of 1,673 subsequent capital injections (again, less than one percent) were in the form of venture capital (Robb et al. 2010). In other words, venture capital financing is a very rare event among the full suite of start-up financing activities.

Since the Kauffman Firm Survey data include a broad cross section of firms, while university spin-outs typically are science- or technology-based, university spin-outs are more likely than the average start-up to be a match for venture capital financing. Still, recent AUTM data show that just 13% of university spin-outs are VC-funded (AUTM 2013). Thus, venture capital remains a minority player in the financing of university spin-outs.

Some states have attempted to provide funding to university technology spin-outs directly. For example, in 2007 the state of Oregon launched a "University Venture Development Fund," which is funded via a generous tax incentive for private donors and which provides proof-of-concept and translational research grants to university spin-outs (Meyer et al. 2011). Similar programs exist at many other universities and in many other states. As Lerner (2009) highlights, however, these efforts can struggle to reconcile multiple and sometimes-competing goals, such as financial returns, company growth, and regional employment.

A second resource requirement, and one that may be closely tied to financing, is facilities and equipment. Academic researchers have offered considerable attention to science parks, business incubators, and other arrangements that attempt to fill early facilities needs for university technology spin-outs (e.g., Drori and Yue 2009; Ferguson and Olofsson 2004; Lindelöf and Löfsten 2003; Link and Link 2003; Link and Scott 2003; Link and Scott 2005; Monck et al. 1988; Siegel, Westhead, and Wright 2003; Westhead 1997; Westhead and Storey 1994; Westhead, Storey, and Cowling 1995). Science parks and other arrangements may provide assistance beyond facilities, too, including networking opportunities, advice, and basic office services. For some technology start-ups, expensive specialized equipment is another important resource (Bania, Eberts, and Fogarty 1993; Smith 1991). Recognizing this fact, some federal grants stipulate that equipment supplied to one organization be made available to other organizations, including companies. (See, for example, the conditions attached to National Cancer Institute Centers of Cancer Nanotechnology Excellence, as reported in Baker 2006.)

Finally, the most important resource for any start-up is people, including potential managers, team members, board members, and advisors (Wright et al. 2007). Here, the academic literature has focused considerable attention on founders, generally ignoring the importance of a deeply skilled and relevant pool of potential employees and advisors. Within this focus on founders, the academic literature has offered special attention to the experiential, psychological, and even biological characteristics associated with the decision to start a company (e.g., Begley and Boyd 1987; Chen, Greene, and Crick 1998; Delmar and Davidsson 2000; Hisrich, Langan-Fox, and Grant 2007; McClelland 1961; Nicolaou et al. 2008; Robinson et al. 1991; Sexton and Bowman 1985; Sexton and Bowman-Upton 1990). One challenge, of course, is that such observations are not readily actionable; they may lead to the conclusion that if a potential entrepreneur does not score above a threshold on a personality quiz, then she should not bother to learn more about entrepreneurship. To preview our remarks on the role of entrepreneurship education, such a perspective fails to recognize the role and purpose of entrepreneurship education, possibly to the detriment of university technology transfer efforts.

Moreover, an emphasis on entrepreneurial characteristics may overshadow other critical features of people and policies. For example, professional investors will readily point to the importance of start-up experience and a longtime horizon for those who wish to engage in entrepreneurship (Byers, Dorf, and Nelson 2014). Saxenian (1994) has noted the importance of employment alternatives in case any one particular start-up fails; if an individual is to pursue a start-up, she needs some assurance that there are other job possibilities in the region given the somewhat high probability of the start-up's failure. Finally, local and regional policies can have an important impact on the availability of people. For example, states that enforce non-compete agreements may experience outward migration of entrepreneurs and inventors, effectively dampening future entrepreneurial activity in the state (Carey 2001; Marx, Strumsky, and Fleming 2009; Samila and Sorenson 2009). The Kauffman Foundation has argued that health care policy has an important effect on entrepreneurship; the loss of employer-tied health insurance can be a disincentive to venture out on one's own (Ortmans 2013).

In considering these different resource requirements, it is critical to recognize that they are all necessary for a university spin-out to meet with success. For example, approaches to university commercialization that only address "gap funding" are almost certain to fail. Similarly, attempts to train CEOs and founders without addressing the larger labor pool of qualified and relevant technical employees are unlikely to succeed. Unless university commercialization efforts via start-ups address *all* of the resource needs outlined above, individual and independent initiatives will have little impact.

This view of basic university spin-out resource requirements—financing, facilities/equipment, and people—grows more complex with the recognition that specific resource needs vary according to specific technologies and markets. Thus, biotechnology, materials, software, and medical-device spin-outs—to name just four examples—require dramatically different sums of money, facilities/equipment, and types of people in order to be successful. (They also differ, of course, in timelines, potential impacts, and other features.) This heterogeneity poses a challenge to university-focused spin-out efforts. Most universities engage in an incredibly wide range of research activities, from biochemistry to materials science to software—and beyond. In turn, attempts to commercialize university research across this wide range of fields and industries must address their very different resource requirements. In most cases, a university or a regional infrastructure simply cannot provide adequate commercialization resources across all of these areas. For example, only a handful of regions in the United States have an extensive labor pool across the full range of industries that university research conducted in that region might impact. In other regions, simply providing general gap funding or entrepreneurship courses for MBAs, as two examples, without addressing these concurrent and diverse resource challenges is unlikely to spur effective commercialization.

Is Entrepreneurship Education the Solution?

As these considerations highlight, commercialization of university research through the creation and successful growth of start-ups is an enormous challenge that hinges on the successful alignment and execution of a variety of resources, activities, goals, and stakeholders. Against this background, entrepreneurship education is a promising means of improving a start-up's viability—though one, as we will argue, that needs to be taken in context and appreciated for its own goals and principles.

Reviewing the considerations in the previous section, it is apparent that entrepreneurship education can have a positive impact upon many of them. For example, educational activities focused on product design and development, prototyping, technology trends, and creativity can help to answer the question of "can you build it?" More critically, they also can suggest whether it makes sense to pursue development in the first place and how entrepreneurs might "pivot" their original idea (Liedtka and Ogilvie 2011; Ries 2011; Seelig 2011). Similarly, a core component of many entrepreneurship curricula is market analysis, including needs assessment, positioning, segmentation, and customer relationship management (Blank and Dorf 2012; Byers, Dorf, and Nelson 2014). Undoubtedly, informed attention to these considerations can positively influence the response to the question of "if you build it, will they come?" Given the shared concern with these questions among both start-ups and established firms who engage in technology commercialization, it is clear that the positive influence of entrepreneurship education is not limited to university spin-outs, but rather extends to all cases of commercialization (and beyond).

Entrepreneurship education can help to mediate other university spin-out challenges, too. Most notably, entrepreneurship education can aid in the development of managers and team members, raising their awareness of potential pitfalls and providing insight into effective strategies and operational activities (Manimala 2008; Rasmussen and Sørheim 2006). Less directly, to the extent that entrepreneurship education is incorporated broadly into the curricula of all students, not only those who wish to start a university spin-out, it can influence policy decisions and broad public support for entrepreneurship. In turn, enlightened public policies can positively impact university spin-outs.

Entrepreneurship education, however, is not a panacea and it cannot be expected to solve every challenge; while it is an important means of improving the effectiveness of university spin-outs, it is not the solution to their challenges in and of itself (McMullan and Long 1987). Moreover, our own experiences as entrepreneurship educators along with considerable scholarly research suggest that entrepreneurship education is most effective when it takes a broad view of its audience, curriculum, and partners. We elaborate on each of these areas below.

Audience

In most universities, the majority of entrepreneurship courses are offered in business schools (Binks, Starkey, and Mahon 2006; Gwynne 2008; Solomon and Fernald 1991. Entrepreneurship as a discipline and as an approach to value creation, however, extends far beyond BBA and MBA programs (Gibb 2002; Hynes 1996; Katz 2003; Ray 1990). In turn, more progressive universities offer courses across a wide range of departments and schools. In the case of university spin-outs, in particular, it is critical to expand entrepreneurship education to engineering and science departments where most of these technologies originate (Meyer et al. 2011; Phan, Siegel, and Wright 2009; World Economic Forum 2009). In our view, this move involves a true integration of entrepreneurship into the broad curricula, not the piecemeal education of a few select students (Thursby, Fuller, and Thursby 2009). Moreover, a truly multidisciplinary approach, which mimics the reality of early-stage technology start-ups, is critical to reflect in entrepreneurship education efforts. This perspective not only involves teaching engineers, scientists, and others about entrepreneurship from a business school perspective, but also suggests that business students, as one example, should become literate in areas of science and technology, among others (Clark 1998; Keogh and Galloway 2004; Menzies 2004; Penaluna and Penaluna 2008). Such integration of disciplines through the lens of entrepreneurship is certain to yield insights that cannot be gleaned through the maintenance of disciplinary silos (Clarysse, Mosey, and Lambrecht 2009; Hill and Kuhns 1994).

Such integration may require logistical and pragmatic adjustments by universities and by individual schools or departments. For example, it should be easy for students to enroll in courses across departments, without facing outside enrollment caps and other barriers that may hinder interdisciplinary engagement. (In turn, such cross-enrollment may force some schools to revisit internal revenue models that may reinforce school boundaries.) Interdisciplinary integration may also require adjustment or special consideration of academic calendars, as when a university's law school runs on semesters but its engineering school runs on quarters.

Finally, it is important that the audience for entrepreneurship education extend beyond students. Faculty, administrators, staff members, and those in the community at large can all benefit from entrepreneurship education and offerings should be tailored accordingly (Siegel and Phan 2005). Such breadth can also reinforce synergies among different groups, as when a faculty member participates in an entrepreneurship program or seminar and subsequently promotes or reinforces the role of entrepreneurship among his or her students.

Curriculum

In other work (Nelson and Byers 2005) we have argued that technology transfer can enhance the entrepreneurship curriculum. For example, figure 5.2 illustrates educational ties between various entrepreneurship and technology transfer groups at Stanford University. Notably, the Office of Technology Licensing (OTL), pictured on the right-hand side of figure 5.2, has ties to every group. These ties reflect cases in which university invention disclosures proved fertile examples for class projects; OTL personnel taught units on intellectual property and licensing; external relationships with companies that had licensed university technologies yielded guest speakers for a course; and other relationships.

At the same time, and consistent with a broad view on entrepreneurship, entrepreneurship education should not be limited to a focus on technology start-ups, which is often the case, but should instead focus on developing perspectives and skills that can applied in many ways across many settings. Thus, entrepreneurship education is not to be confused with conducting a feasibility study, writing a business plan, or participating in business plan competitions. While such activities can be valuable academic exercises, they can also lead students to confuse endless analysis, number crunching, and polished "suits and slides" with the actual work of starting and managing an organization. Moreover, they may imply that entrepreneurship can be reduced to "picking" an idea and following a prescribed set of steps according to a specific timeline as outlined in a syllabus. Instead, like all curricular efforts, feasibility studies, business plans, and business plan competitions must be placed in context and must first and foremost be approached as educational, not company- or economy-building, activities.

Indeed, emerging evidence indicates that entrepreneurship education should take a variety of forms, including traditional courses, work-study programs, internships, mentoring relationships, workshops, seminars, and all-campus initiatives such as "Entrepreneurship Week," a worldwide event centered on hands-on activities that expose college students and others to creative problem solving and other aspects of entrepreneurship (Fayolle and Gailly 2008; Garavan and O'Cinneide 1994; Pittaway and Cope 2007; Rutger 2008; Wee and Lynda 2004). Within these structures and events, entrepreneurship education might include games, simulations, case studies, feasibility studies, discussion of readings, lectures, interviews, field studies, hands-on exercises, and other activities

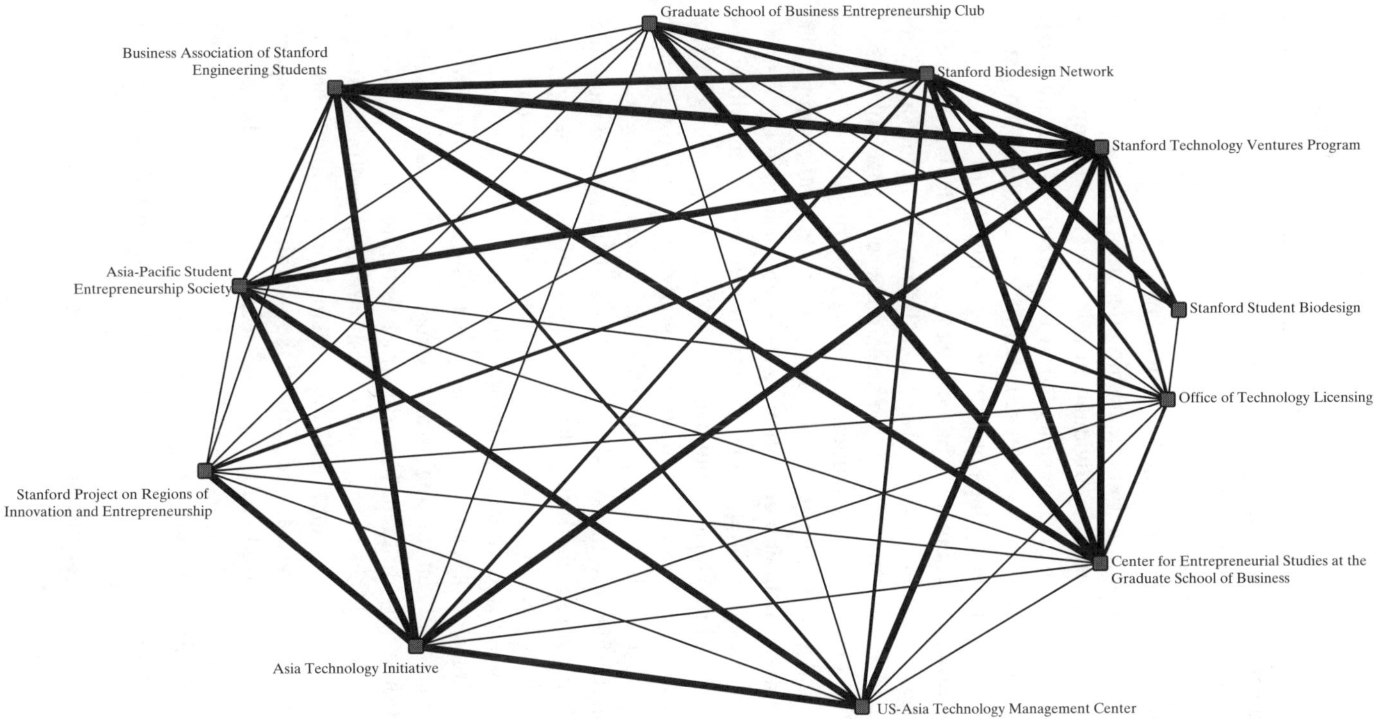

FIGURE 5.2. Educational ties between technology transfer and entrepreneurship groups at Stanford University. Ties indicate responses to the question, "Faculty or staff from your group are involved in teaching students from." Thickness of line indicates frequency on a five point scale from "never" (which has no line) to "nearly always" (which has the thickest line). Image is from Nelson and Byers, 2005.

(Clarysse, Mosey, and Lambrecht 2009; DeTienne and Chandler 2004; Katz 1999; Kuratko 2005; Mustar 2009; Verzat, Byrne, and Fayolle 2009). Such a diversity of offerings permits students to engage with entrepreneurship in different ways and with different levels of commitment, facilitating low-risk experimentation along with content and formats tailored to specific interests (Vesper and McMullan 1988).

In terms of content, entrepreneurship education plays an important role not only in informing individuals how to start a company, but also in teaching them how to manage and grow organizations (Klandt 2004). Mullins and Komisar (2009) offer compelling evidence that most start-ups will switch business plans or even entire industries over the course of their growth. For technology-based start-ups, in particular, such shifts may be the norm rather than the exception (Ries 2011). In the same vein, Collins and Lazier (1995) focus on the unique challenges of managing small- to mid-sized companies, exploring the crucial steps between "launch" and a large stable company. These works and others imply that limiting entrepreneurship education to the point of launch or raising a first round of outside funding may be akin to demonstrating tilling and planting to would-be farmers but neglecting to discuss watering, fertilizing, harvesting, crop rotation, and other equally important topics. Fully executing on this "lifecycle" approach to entrepreneurship education also suggests that entrepreneurship should be fully integrated into other courses that address challenges that typically confront later-stage organizations. A number of universities (e.g., Stanford and MIT), in fact, extend the content and conceptualization of entrepreneurship even further, considering it a leadership training initiative rather than an area focused specifically on starting organizations.

Given the many different fields in which entrepreneurs act, courses and content should also reflect upon, compare, and contrast different settings, rather than attempting to apply a single model to all entrepreneurial endeavors. Thus, as one example, some universities offer special courses on social entrepreneurship, recognizing its unique characteristics (Smith et al. 2008; Tracey and Phillips 2007). In the area of technology commercialization, entrepreneurship courses must be careful to acknowledge the important distinctions between different fields and different types of technologies, again recognizing the significant heterogeneity between cases of university technology spin-outs rather than attempting to treat them equally.

Finally, the impacts of entrepreneurship education must be perceived as lifelong and not immediately discernible. Even for those students who will start a company, the vast majority are not in a position to be effective leaders immediately upon graduation and they are better served by first gaining relevant experience. Entrepreneurship education may have an immediate impact on their perceptions of the feasibility and desirability of starting an organization (Peterman and Kennedy 2003) and on their skills (Souitaris, Zerbinati, and Al-Laham 2007). But, Kauffman Foundation data show that technology-based company founders rarely are fresh graduates, since the median age is 39 and many founders are much older (Wadhwa, Freeman, and Rissing 2008).

Moreover, many students may never spin out a technology or start an organization. They may still carry entrepreneurial skills and attitudes into existing organizations, however, finding that they lead to enhanced problem solving and development of new initiatives, along with overall increases in job satisfaction and performance (Hindle and Cutting 2002). In fact, the realization on the part of some students that they should *not* start an organization must be regarded as a positive outcome. Still other students may find that the primary role of entrepreneurship education is in better understanding the entrepreneurial process and, thereby, developing more accurate and impactful policies or research agendas (Klandt 2004). These outcomes, too, should be regarded as successes even though they do not involve starting an organization.

Together, these broad perspectives on the role and impacts of entrepreneurship education suggest that university administrators and others should not expect to see the effects of entrepreneurship education quickly nor through simple counts of the number of spin-outs. Diverse goals demand diverse measures.

External Engagement

Finally, entrepreneurship education should not be considered the purview or responsibility of universities alone. Instead, educational institutions, non-profits, firms, and government organizations at all levels should play an active role in developing and supporting educational initiatives. For example, as they develop and execute technology-focused entrepreneurship education activities, universities might partner with existing technology-based firms. Such engagement can increase the resource base and ensure that curricula strike an appropriate balance between academic theory and practice (Collins, Smith, and Hannon 2006;

Kuratko 2005; Roebuck and Brawley 1996). Since the commercialization of university technologies is dependent not only upon actions by universities but also by a number of external stakeholders, engagement of these stakeholders in educational efforts can also build relationships that can be leveraged in commercialization efforts (Todorovic and Sontornpithug 2008). In fact, universities are uniquely positioned to serve as a hub for both innovation and educational activities, drawing together diverse groups across a region and/or industry (Clark 1998; World Economic Forum 2009). Viewed in this way, the efforts surrounding activities such as Startup Weekend (Nager, Nelsen, and Nouyrigat 2012) should be viewed as complementary to internal university efforts.

Reconsidering the Relationship between Entrepreneurship Education and Technology Transfer

As evident in our comments above, many effective practices and approaches in entrepreneurship education can enhance technology transfer efforts, though entrepreneurship education also must be recognized as distinct in its goals, orientation, approach, and audience. Just as the realm of technology transfer extends beyond entrepreneurship, the realm of entrepreneurship involves more than technology transfer initiatives. Table 5.1 highlights the differences between technology transfer and entrepreneurship education along a number of dimensions.

As the table indicates, technology transfer is focused on economic and technological outcomes whereas entrepreneurship education is focused on educational development, relationships, and outcomes. Technology transfer, therefore, maintains a strong commercial orientation, while entrepreneurship education exhibits a more nuanced relationship with commercial goals according to the specific perspective at play. Technology transfer also adheres to a shorter timeline than entrepreneurship education: most licenses for new technologies are signed within a year of disclosure (Elfenbein 2006) and patent terms limit revenues to an absolute maximum of 20 years. Thus, Markman et al. (2005) argue that most university technology transfer offices are risk-averse and focused on short-term cash maximization. By contrast, entrepreneurship education takes a long-term perspective and aims for a lifelong impact.

Assessment of technology transfer and entrepreneurship education also differs. Observers typically evaluate technology transfer according to relatively straightforward metrics such as licenses signed or spin-outs

TABLE 5.1. **Distinctions between university technology transfer and entrepreneurship education (adapted from Nelson and Byers 2005).**

	Technology Transfer	Entrepreneurship Education
Goals and Mission	Commercialize inventions; generate income; contribute to regional economic development; spur start-ups	Develop leadership skills; integrate courses and disciplines; provide the foundation for new businesses; forge links between academic and business communities
Commercial Orientation	Significant	Mixed—prevalent in traditional start-ups, but less apparent in social entrepreneurship and other non-market-focused entrepreneurship activities
Time Horizon	Short-term	Long-term
Assessment	Inventions commercialized; licenses executed; revenue; start-ups; regional employment	Attitudes, behaviors, and capabilities, including creativity, risk-taking, pursuit of opportunities, self-confidence, self-efficacy, and organizational founding
Providers and Constituency	Administrators focused on corporate relations and university intellectual property; faculty and students with inventions that hold commercial potential; licensees and potential licensees, including start-ups, small companies, and large companies	Faculty, students, entrepreneurs and other members of the entrepreneurship ecosystem, representing a wide range of activities and disciplines

generated.[1] By contrast, the set of outcomes for entrepreneurship education efforts and, therefore, the assessment of these efforts is very broad. As a result, measurement and assessment of the effects of entrepreneurship education remains an important challenge (Falk and Alberti 2000; Fayolle, Gailly, and Lassas-Clerc 2006; Menzies and Paradi 2003; Peterman and Kennedy 2003).

Finally, the providers and constituencies differ for technology transfer versus entrepreneurship education. Technology transfer engages only those faculty and students with an invention of commercial potential, and it strives to link them with external licensors, primarily serving the interests of large companies (Siegel, Waldman, and Link 2003; Siegel et al. 2004). By contrast, entrepreneurship education engages a much wider cross section of the university population, and it strives to engage a much broader cross section of external partners.

Together, these many differences signal that while technology transfer and entrepreneurship education may be complementary, they also must be recognized as very different activities. Absent such recognition, entrepreneurship education programs run the risk of missing out on broader opportunities beyond technology transfer or, in the worst case, adopting inappropriate goals, metrics, and timelines that fail to maintain the educational mission of entrepreneurship education as central (Meyer et al. 2011; Nelson and Byers 2005).

Given the idiosyncratic history of many entrepreneurship programs, many educators have not explicitly considered their program's relationship to technology transfer, including the extent to which this orientation implicitly or explicitly affects program offerings, goals, and constituencies. As such, we propose the following set of questions by which one can assess this relationship for a particular university or program:

1. What proportion of entrepreneurship students is within the business school versus outside of it? Is the entrepreneurship "student body" dominated by a single discipline?
2. Do we offer entrepreneurship courses beyond those courses that train for the commercialization of technologies? To what extent does the entrepreneurship curriculum emphasize topics beyond technology commercialization?
3. Do we offer a wide range of courses in a wide range of formats (e.g., seminars, speaker series, project-based courses, work-study programs, etc.)?
4. Do our entrepreneurship instructors represent a wide range of backgrounds and experiences? Do we look for instructional talent beyond those individuals whose experience is based on commercialization of technologies?
5. How many different units across our university are actively engaged in entrepreneurship education efforts? What is the extent of their engagement and how balanced is it across units?
6. What are the perceptions of the average student on campus about entrepreneurship? How do these perceptions align with our program's mission and goals?
7. What expectations do senior administrators, alumni, funders, and the gen-

eral public have regarding our entrepreneurship efforts? To what extent is their focus on commercialization versus education?

8. What are the sources of funding for our entrepreneurship education efforts? Are funding sources (including appropriations and grants) independent of technology commercialization links and programs?
9. What timeline do students, faculty, and others attach to entrepreneurship? Do they take a long-term perspective?
10. What evidence do we have of the success of our entrepreneurship education efforts? What metrics do we implicitly or explicitly emphasize when considering our entrepreneurship program's success? How important is commercial impact in assessing our entrepreneurship center? What metrics or statistics do we highlight when describing our program?

There is no answer key for these questions. Rather, they are intended to highlight implicit assumptions about the role of entrepreneurship education and its relationship to technology transfer. Those respondents whose answers heavily emphasize technology transfer may find that their entrepreneurship education efforts could be even more impactful by moving beyond this particular conceptualization. Conversely, those respondents who find too little engagement with technology transfer may find that their entrepreneurship education efforts are failing to address this important area and to take advantage of complementarities with it.

More generally, the fact that entrepreneurship education and technology transfer are complementary yet also distinct highlights a crucial challenge: how to facilitate synergies between the activities while not allowing one to co-opt the other. In other work, we have proposed that the concept of "organizational modularity" offers a promising model (Nelson and Byers 2005). In a seminal article, Weick (1976) argued that when an organization pursues multiple goals that may conflict, its formal structure may be only "loosely" integrated. Adkison (1979) offered an early application of Weick's concept through her study of the Kansas Public School System. She found that "loose coupling" between participants allowed them to pursue unique roles and responsibilities while avoiding conflict. Tushman and O'Reilly (2004) drew upon these same ideas in developing their concept of "ambidextrous" organization. In their view, organizations that attempt to apply a single model or perspective to all subunits realize poor outcomes compared to those organizations that recognize and facilitate differences. (See also Martin and Eisenhardt 2003.)

At the same time, organizations profit from ensuring that these units coordinate activities and initiatives where synergies exist, as when entrepreneurship education and technology transfer are mutually beneficial. This coordination depends first and foremost upon awareness among various members about the activities of others. For example, Tushman and O'Reilly (2004) highlighted the benefits from integrated top management teams when units are independent, since this integration facilitates awareness and coordination across the independent units. In a study of twelve cross-business synergy initiatives, Martin and Eisenhardt (2003) found that high-performing initiatives originated in the business units, not at the corporate level, and that high-performing initiatives had an "engaged multi-business team decision process," rather than a top-down corporate decision process. Similarly, Tsai's (2002) investigation of a large diversified organization revealed that formal hierarchical structure had a negative effect on knowledge sharing between units, while informal lateral relations had a positive effect.

Together, these studies, and others, suggest that awareness and cooperation function best when allowed to emerge from the bottom up. For example, in a comparison of university technology transfer performance in the U.S. versus Sweden, Goldfarb and Henrekson (2003) found that much of the higher performance in the U.S. was attributable to its bottom-up approach versus Sweden's top-down approach. Similarly, those universities that are attempting to make the most of relationships between entrepreneurship education and technology transfer must, somewhat ironically, avoid planning these relationships in a centralized fashion. Instead, relations should emerge organically, with administrators providing some resources and facilitating connections, but not driving policies and initiatives. At the same time, individual participants in the entrepreneurship and technology transfer ecosystems must take pains to ensure that their bottom-up efforts do not simply include the "usual suspects" and reinforce existing relationships; the nurturing and growth of entrepreneurship education and technology transfer demand the constant infusion of new ideas, participants, and programs.

Summary and Conclusion

Both entrepreneurship education and university technology transfer have witnessed dramatic increases over the past two decades (AUTM

2013; Charney and Libecap 2000; Katz 2003; Mowery et al. 2001; Solomon, Duffy, and Tarabishy 2002; Vesper and Gartner 1997). While there are a number of dimensions to each of these areas, their simultaneous growth reflects, in part, the many complementarities between them. On a fundamental level, the commercialization of university technologies requires vision, leadership, persistence, imagination, and the ability to assemble critical resources, including financial and human capital. Entrepreneurship education strives to develop these very skills and capabilities. As a result, the close relationship between these areas should not come as a surprise.

Our review of the entrepreneurship education and technology transfer literatures highlighted a number of ways in which the two fields can inform one another and in which, on a pragmatic level, programs and resources may be shared and integrated. Specifically, we discussed how broad perspectives on both the audience and curriculum for entrepreneurship education, along with deep engagement of external partners, can enhance technology commercialization and education efforts overall. For those entrepreneurship programs that have not engaged with university technology transfer personnel and programs, such engagement represents a straightforward and effective means of extending the impact and "real world" engagement of their efforts.

At the same time, we described a number of dimensions along which entrepreneurship education and technology transfer differ. A major challenge lies in ensuring that these differences are respected and maintained. For example, a program director interviewed by Meyer et al. (2011: 189) acknowledged that

> many programs similar to ours started out with education as the central goal (as we did), but through mission creep, educational goals gave way to trying to maximize revenue, deals, IP licenses, business competition prizes, and other metrics. If we want to avoid pitfalls that have diverted other programs, we must never lose sight of the importance of delivering value to all of our students.

The siren song of start-ups, management titles, prize money, and investors can be alluring to students, faculty, and administrators alike. Focusing entrepreneurship education efforts primarily upon technology commercialization efforts, however, sacrifices the incredible breadth of the field for a very high-risk low-probability outcome with a limited educational impact. By contrast, by *not* focusing primarily on technology transfer and instead ensuring that entrepreneurship education maintains a wide set of goals and a diverse audience, universities can effectively strengthen their entrepreneurial ecosystem—yielding, ironically, even greater long-term benefits for technology transfer.

Note

1. Siegel, Waldman, and Link (2003) note that other important aspects of technology transfer offices, such as organizational practices, are not amenable to quantification and, therefore, are often overlooked.

References

Acs, Z. J., D. B. Audretsch, and M. P. Feldman. 1994. "R & D Spillovers and Recipient Firm Size." *Review of Economics and Statistics* 76 (2): 336–40.

Adkison, J. A. 1979. "Conflict and stress in interorganizational structures." ERIC Document Reproduction Service No. ED 170861.

Agrawal, A., and R. Henderson. 2002. "Putting patents in context: Exploring knowledge transfer from MIT." *Management Science* 48 (1): 44–60.

Antoncic, B., and R. D. Hisrich. 2001. "Intrapreneurship: Construct refinement and cross-cultural validation." *Journal of Business Venturing* 16 (5): 495–527.

Antoncic, B., and R. D. Hisrich. 2003. "Clarifying the intrapreneurship concept." *Journal of Small Business and Enterprise Development* 10 (1): 7–24.

Association of University Technology Managers (AUTM). 2013. 2012 U.S. Licensing Survey Summary. Northbrook, IL: AUTM.

Audretsch, D. B., M. C. Keilbach, and E. Lehmann. 2006. "Entrepreneurship and economic growth." Oxford: Oxford University Press.

Audretsch D., and P. Stephan. 1996. "Company-scientist locational links: The case of biotechnology." *American Economic Review* 86: 641–52.

Baker, M. 2006. *National Cancer Institute Funds New Nanotechnology Center.* Stanford Report, March 1, 2006

Bania, N., R. W. Eberts, and M. S. Fogarty. 1993. "Universities and the Startup of New Companies: Can We Generalize from Route 128 and Silicon Valley?" *Review of Economics and Statistics* 75 (4): 761–66.

Baum, J. A. C., and B. S. Silverman. 2004. "Picking winners or building them? Alliance, intellectual, and human capital as selection criteria in venture financing and performance of biotechnology startups." *Journal of Business Venturing* 19 (3): 411–36.

Béchard, J. P., and D. Grégoire. 2005. "Entrepreneurship education research revisited: The case of higher education." *Academy of Management Learning and Education* 4 (1): 22–43.

Begley, T., and D. Boyd. 1987. "Psychological characteristics associated with performance in entrepreneurial firms and smaller businesses." *Journal of Business Venturing* 2: 79–93.

Benneworth, P., and D. Charles. 2005. "University spin-off policies and economic development in Less successful regions: Learning from two decades of policy practice." *Journal of European Planning Studies* 13 (4): 537–57.

Binks, M., K. Starkey, and C. Mahon. 2006. "Entrepreneurship education and the business school." *Technology Analysis and Strategic Management* 18 (1): 1–18.

Blank, S., and B. Dorf. 2012. *The Startup Owner's Manual*. Pescadero, CA: K&S Ranch Publishers.

Byers, T., R. C. Dorf, and A. J. Nelson. 2014. *Technology Ventures: From Idea to Enterprise*. 4th ed. New York: McGraw-Hill.

Carey, K. L. 2001. "Beyond the Route 128 paradigm: Emerging legal alternatives to the non-compete agreement and their potential effect on developing high-technology markets." *J. Small and Emerging Bus. L.* 5 (1): 135–60.

Carrier, C. 1993. "Intrapraneurship in Large Firms and SMEs: A Comparative Study." *International Small Business Journal* 12 (3): 54–61.

Charney, A., and G. Libecap. 2000. "The Impact of Entrepreneurship Education: An Evaluation of the Berger Entrepreneurship Program at the University of Arizona, 1985–1999." Revised final report to the Kauffman Center for Entrepreneurial Leadership. Kansas City: Ewing Marion Kauffman Foundation, November 29.

Chen, C., P. Greene, and A. Crick. 1998. "Does entrepreneurial self-efficacy distinguish entrepreneurs from managers?" *Journal of Business Venturing* 13: 295–316.

Christensen, C. M. 1997. *The Innovator's Dilemma: When New Technologies Cause Great Firms to Fail*. Cambridge, MA: Harvard Business School Press.

Clark, B. R. 1998. *Creating Entrepreneurial Universities: Organizational Pathways of Transformation. Issues in Higher Education*. New York: Elsevier.

Clarysse, B., S. Mosey, and I. Lambrecht. 2009. "New Trends in Technology Management Education: A View From Europe." *Academy of Management Learning and Education* 8 (3): 427–43.

Collins, J. C., and W. C. Lazier. 1995. *Managing the Small to Mid-Sized Company: Concepts and Cases*. Chicago: Irwin.

Collins, L. A., A. J. Smith, and P. D. Hannon. 2006. "Applying a Synergistic Learning Approach in Entrepreneurship Education." *Management Learning* 37 (3): 335–54.

Degroof, J. J., E. B. Roberts. 2004. "Overcoming Weak Entrepreneurial Infrastructure for Academic Spin-Off Ventures." *Journal of Technology Transfer* 29(3–4): 327–57.

Delmar, F. and P. Davidsson. 2000. "Where do they come from? Prevalence and characteristics of nascent entrepreneurs." *Entrepreneurship and Regional Development* 12 (1): 1–23.

DeTienne, D. R., and G. N. Chandler. 2004. "Opportunity Identification and Its Role in the Entrepreneurial Classroom: A Pedagogical Approach and Empirical Test." *Academy of Management Learning and Education* 3 (3): 242–57.

Di Gregorio, D., and S. Shane. 2003. "Why Do Some Universities Generate More Start-Ups than Others?" *Research Policy* 32: 209–27.

Drori, G. S., and Y. Yue. 2009. "The innovation centre: A global model for entrepreneurship in the era of globalization." *International Journal of Entrepreneurship and Innovation* 10 (3): 171–80.

Elfenbein, D. W. 2006. "Publications, patents, and the market for university inventions." *Journal of Economic Behavior and Organization* 63 (4): 688–715.

Falk, J., and F. Alberti. 2000. "The assessment of entrepreneurship education." *Industry and Higher Education* 14 (2): 101–8.

Fayolle, A., and B. Gailly. 2008. "From craft to science: Teaching models and learning processes in entrepreneurship education." *Journal of European Industrial Training* 32 (7): 569–93.

Fayolle, A., B. Gailly, and N. Lassas-Clerc. 2006. "Assessing the impact of entrepreneurship education programs: A new methodology." *Journal of European Industrial Training* 30 (8–9): 701–20.

Feldman, M., and P. Desrochers. 2003. "Research Universities and Local Economic Development: Lessons from the History of the Johns Hopkins University." *Industry and Innovation*.10 (1): 5–24.

Ferguson, R., and C. Olofsson 2004. "Science Park and the Development of New Technology-Based Firms: Location, Survival and Growth." *Journal of Technology Transfer* 29 (1): 5–17.

Franklin, S. J., M. Wright, and A. Lockett. 2001. "Academic and surrogate entrepreneurs in university spin-out companies." *Journal of Technology Transfer* 26(1–2):127–41.

Garavan, T. N., and B. O'Cinneide. 1994. "Entrepreneurship Education and Training Programmes: A Review and Evaluation (Part 1)." *Journal of European Industrial Training* 18 (8): 3–12.

Gibb, A. 2002. "In pursuit of a new 'enterprise' and 'entrepreneurship' paradigm for learning: creative destruction, new values, new ways of doing things and new combinations of knowledge." *International Journal of Management Reviews* 4 (3): 213–31.

Goldfarb, B., and M. Henrekson. 2003. Bottom-up versus top-down policies towards the commercialization of university intellectual property." *Research Policy* 32 (4): 639–58.

Gwynne, P. 2008. "More Schools Teaching Entrepreneurship." *Research Technology Management* 51 (2): 6–8.

Hallen, B. L. 2008. "The causes and consequences of the initial network positions of new organizations: From whom do entrepreneurs receive investments?" *Administrative Science Quarterly* 53 (4): 685–718.

Hart, D. M., ed. 2003. "The emergence of entrepreneurship policy: governance, start-ups, and growth in the U.S. knowledge economy." Cambridge: Cambridge University Press.

Henderson, R. M., and K. B. Clark. 1990. "Architectural innovation: The reconfiguration of existing product technologies and the failure of established firms." *Administrative Science Quarterly* 35 (1): 9–30.

Hill, R. C., and B. A. Kuhns. 1994. "Experiential learning through cross-campus cooperation: Simulating and initiating technology. . . . *Simulation and Gaming* 25 (3): 368.

Hindle, K., and N. Cutting. 2002. "Can Applied Entrepreneurship Education Enhance Job Satisfaction and Financial Performance? An Empirical Investigation in the Australian Pharmacy Profession." *Journal of Small Business Management* 40 (2): 162–67.

Hisrich, R., J. Langan-Fox, and S. Grant. 2007. "Entrepreneurship Research and Practice: A Call to Action for Psychology." *American Psychologist* 62 (6): 575–89.

Hynes, B. 1996. "Entrepreneurship education and training: Introducing entrepreneurship into non-business disciplines." *Journal of European Industrial Training* 20 (8): 10–17.

Jeng, L. A., P. C. Wells. 2000. "The determinants of venture capital funding: evidence across countries." *Journal of Corporate Finance* 6 (3): 241–89.

Kaplan, S., and P. Strömberg. 2004. "Characteristics, Contracts, and Actions: Evidence from Venture Capitalist Analyses." *Journal of Finance* 59 (5): 2177–210.

Katz, J. A. 1999. "Institutionalizing Elegance: When Simulation Becomes a Requirement." *Simulation and Gaming* 30 (3): 332–36.

Katz, J. A. 2003. "The chronology and intellectual trajectory of American entrepreneurship education: 1876–1999." *Journal of Business Venturing* 18 (2): 283–300 .

Keogh, W., and L. Galloway. 2004. "Teaching enterprise in vocational disciplines: Reflection on positive experience." *Management Decision* 42 (3/4): 531–41.

Klandt, H. 2004. "Entrepreneurship Education and Research in German-Speaking Europe." *Academy of Management Learning and Education* 3 (3): 293–301.

Kortum, S., and J. Lerner. 2000. "Assessing the contribution of venture capital to innovation." *RAND Journal of Economics* 31 (4): 674–92.

Kuratko, D. F. 2005. "The Emergence of Entrepreneurship Education: Development, Trends, and Challenges." *Entrepreneurship: Theory and Practice* 29 (5): 577–97.

Lerner, J. 2005. "The University and the Start-Up: Lessons from the Past Two Decades." *Journal of Technology Transfer* 30 (1–2): 49–56.

Lerner, J. 2009. *Boulevard of Broken Dreams: Why Public Efforts to Boost Entrepreneurship and Venture Capital Have Failed—and What to Do about It.* Princeton, NJ: Princeton University Press.

Liedtka, J., and T. Ogilvie. 2011. *Designing for Growth: A Design Thinking Tool Kit for Managers.* New York: Columbia Business School Publishing.

Lindelöf, P., and H. Löfsten. 2003. "Science Park Location and New Technology-Based Firms in Sweden: Implications for Strategy and Performance." *Small Business Economics* 20 (3): 245–58.

Link, A. N., and K. R. Link. 2003. On the Growth of U.S. Science Parks. *Journal of Technology Transfer* 28: 81–85.

Link, A. N., and J. T. Scott. 2003. "Science Parks and the Generation of University-Based Knowledge: An Exploratory Study." *International Journal of Industrial Organization* 21 (9): 1323–56.

Link, A. N., and J. T. Scott. 2005. "Opening the Ivory Tower's Door: An Analysis of the Determinants of the Formation of U.S. University Spin-Off Companies." *Research Policy* 34: 1106–12.

Lockett, A., and M. Wright. 2005. "Resources, capabilities, risk capital and the creation of university spin-out companies." *Research Policy* 34 (7): 1043–57.

Lockett, A., M. Wright, and S. Franklin. 2003. "Technology transfer and universities' spin-out strategies." *Small Business Economics* 20 (2): 185–200.

Mair, J., and I. Marti. 2006. "Social entrepreneurship research: A source of explanation, prediction, and delight." *Journal of World Business* 41 (1): 36–44.

Manimala, M. J. 2008. "Entrepreneurship education in India: An assessment of SME training needs against current practices." *International Journal of Entrepreneurship and Innovation* 8 (6): 624–47.

Markman, G. D., et al. 2005. "Entrepreneurship and university-based technology transfer." *Journal of Business Venturing* 20 (2): 241–63.

Martin, J. A., and K. M. Eisenhardt. 2003. "Cross-business synergy: Recombination, modularity, and the multi-business team." Academy of Management 2003 Best Paper Proceedings, Business Policy and Strategy Division.

Martin, R., and S. Osberg. 2007. "Social Entrepreneurship: The Case for Definition." *Stanford Social Innovation Review* (Spring): 28–39.

Marx, M., D. Strumsky, and L. Fleming. 2009. "Mobility, Skills, and the Michigan Non-Compete Experiment." Management Science. 55: 875–89.

McClelland, D. 1961. *The Achieving Society*. Princeton, NJ: Van Nostrand.

McMullan, W. E., and W. A. Long. 1987. "Entrepreneurship Education in the Nineties." *Journal of Business Venturing* 2 (3): 261–75.

Menzies, T. V. 2004. "Are universities playing a role in nurturing and developing high-technology entrepreneurs? The administrators' perspective." *International Journal of Entrepreneurship and Innovation* 5 (3): 149–57.

Menzies, T. V., and J. C. Paradi. 2003. "Entrepreneurship education and engineering students—Career path and business performance." *International Journal of Entrepreneurship and Innovation* 4 (2): 121–32.

Meyer, A. D., et al. 2011. "Creating a university technology commercialisation program: Confronting conflicts between learning, discovery and commercialization goals." *International Journal of Entrepreneurship and Innovation Management* 13 (2): 179–98.

Monck, C. S. P., et al. 1988. *Science Parks and the Growth of High Technology Firms*. London: Croom Helm.

Mowery, D. C., et al. 2001. "The Growth of Patenting and Licensing by U.S. Universities: An Assessment of the Effects of the Bayh-Dole Act of 1980." *Research Policy* 30: 99–119.

Mueller, P. 2006." Exploring the knowledge filter: How entrepreneurship and university-industry relationships drive economic growth." *Research Policy* 35 (10): 1499–508.

Mullins, J., and R. Komisar. 2009. *Getting to Plan B: Breaking Through to a Better Business Model*. Cambridge, MA: Harvard University Press.

Mustar, P. 2009. "Technology Management Education: Innovation and Entrepreneurship at MINES ParisTech, a Leading French Engineering School." *Academy of Management Learning and Education* 8 (3): 418–42.

Nager, M., C. Nelsen, and F. Nouyrigat. 2012. *Startup Weekend: How to Take a Company from Concept to Creation in 54 Hours*. Hoboken, NJ: John Wiley and Sons.

Nelson, A. J. 2012. "Putting university research in context: Assessing alternative measures of production and diffusion at Stanford." *Research Policy* 41 (4): 678–91.

Nelson, A. J., and T. Byers. 2005. "Organizational Modularity and Intra-University Relationships Between Entrepreneurship Education and Technology Transfer." In *University Entrepreneurship and Technology Transfer: Process, Design, and Intellectual Property*, edited by Gary Libecap, 275–311. Stamford, CT: Elsevier Science/JAI Press.

Nerkar, A., and S. Shane. 2003. "When Do Startups that Exploit Academic Knowledge Survive?" *International Journal of Industrial Organization* 21 (9): 1391–410.

Nicolaou, N., et al. 2008. "Is the Tendency to Engage in Entrepreneurship Genetic?" *Management Science* 54 (1): 167–79.

Oden, H.W. 1997. *Managing Corporate Culture, Innovation, and Intrapreneurship*. Westport, CT: Quorum Books.

Ortmans, J. 2013. No Absolutes in the Affordable Care Act. http://www.entrepreneurship.org/Blogs/Policy-Forum-Blog/2013/July/No-Absolutes-in-the-Affordable-Care-Act.aspx Accessed 24 October 2013.

O'Shea, R., et al. 2005. "Entrepreneurial orientation, technology transfer and spinoff performance of US universities." *Research Policy* 34 (7): 994–1009.

Penaluna, A., and K. Penaluna. 2008. "Business Paradigms in Einstellung: Harnessing Creative Mindsets, A Creative Industries Perspective." *Journal of Small Business and Entrepreneurship* 21 (2): 231–50.

Peterman, N. E., and J. Kennedy. 2003. "Enterprise Education: Influencing Students' Perceptions of Entrepreneurship." *Entrepreneurship: Theory and Practice* 28 (2): 129–44.

Phan, P., D. Siegel, and M. Wright. 2009. "New Developments in Technology Management Education: Background Issues, Program Initiatives, and a Research Agenda." *Academy of Management Learning and Education* 8 (3): 324–36.

Pinchot, G. 1985. *Intrapreneuring: Why You Don't Have to Leave the Corporation to Become an Entrepreneur*. New York: Harper Collins.

Pittaway, L., and J. Cope. 2007. "Simulating Entrepreneurial Learning." *Management Learning* 38 (2): 211–33.

Rasmussen, E.A., and R. Sørheim. 2006. "Action-based entrepreneurship education." *Technovation* 26 (2): 185–94.

Ray, D. 1990. "Liberal arts for entrepreneurs." *Entrepreneurship: Theory and Practice* 15 (2): 79–93.

Ries, E. 2011. *The Lean Startup*. New York: Crown Business.

Robb, A., et al. 2010. *An Overview of the Kauffman Firm Survey: Results from the 2004–2008 data*. Kansas City, MO: Ewing Marion Kauffman Foundation.

Robinson, P., et al. 1991. "An Attitude Approach to the Prediction of Entrepreneurship." *Entrepreneurship: Theory and Practice* 15(4):13–31.

Roebuck, D. B., and D. E. Brawley. 1996. "Forging links between the academic and business communities." *Journal of Education for Business* 71 (3): 125–28.

Rosenberg, N., 1996. "Uncertainty and technological change." In *The Mosaic of Economic Growth*, edited by R. Landau, T. Taylor, and G. Wright, 334–56. Stanford, CA: Stanford University Press.

Rutger, H. 2008. *Week Devoted to Innovation, Can-Do Attitude Returns to Campus*. Stanford Report. February 20.

Samila, S., and O. Sorenson. 2009. *Non-Compete Covenants: Incentives to Innovate or Impediments to Growth (May 28, 2009)*. Available at SSRN: http://ssrn.com/abstract=1411172.

Saxenian, A. 1994. *Regional Advantage: Culture and Competition in Silicon Valley and Route 128*. Cambridge, MA: Harvard University Press.

Seelig, T. 2011. *inGenius: A Crash Course on Creativity*. New York: HarperOne.

Sexton, D. L., and N. Bowman. 1985. "The entrepreneur: A capable executive and more." *Journal of Business Venturing* 1 (1): 129–40.

Sexton, D. L., and N. Bowman-Upson. 1990. "Female and male entrepreneurs: Psychological characteristics and their role in gender-related discrimination." *Journal of Business Venturing* 5 (1): 29–36.

Shane, S., and S. Venkataraman. 2000. "The promise of entrepreneurship as a field of research." *Academy of Management Review* 25 (1): 217–26.

Shepherd, D. A., E. J. Douglas, and M. Shanley. 2000. "New venture survival: Ignorance, external shocks, and risk reduction strategies." *Journal of Business Venturing* 15 (5–6): 393–410.

Siegel, D. S., and Phan, P. H. 2005. "Analyzing the Effectiveness of University Technology Transfer: Implications for Entrepreneurship Education." *Advances in the Study of Entrepreneurship, Innovation and Economic Growth* 16: 1–38.

Siegel, D. S., D. Waldman, and A. Link. 2003. "Assessing the impact of organizational practices on the relative productivity of university technology transfer offices: An exploratory study." *Research Policy* 32:127–48.

Siegel, D. S., et al. 2004. "Toward a model of the effective transfer of scientific knowledge from academicians to practitioners: Qualitative evidence from the commercialization of university technologies." *Journal of Engineering and Technology Management* 21 (1–2): 115–42.

Siegel, D. S., P. Westhead, and M. Wright. 2003. "Assessing the Impact of Science Parks on the Research Productivity of Firms: Exploratory Evidence from the United Kingdom." *International Journal of Industrial Organization* 21 (9): 1357–69.

Siegel, D. S., M. Wright, and A. Lockett. 2007. "The rise of entrepreneurial activity at universities: Organizational and societal implications." *Industrial and Corporate Change* 16 (4): 489–504.

Smilor, D. V. Gibson, and G. B. Dietrich. 1990. "University spin-out companies: Technology start-ups from UT–Austin." *Journal of Business Venturing* 5 (1): 63–76.

Smith, B. R., et al. 2008. "Social entrepreneurship: A grounded learning approach to social value creation." *Journal of Enterprising Culture* 16 (4): 339–62.

Smith, H. L. 1991. "The role of incubators in local industrial development: The cryogenics industry in Oxfordshire." *Entrepreneurship and Regional Development* 3 (2): 175–94.

Solomon, G. T., S. Duffy, and A. Tarabishy. 2002. "The state of entrepreneurship education in the United States: A nationwide survey and analysis." *International Journal of Entrepreneurship Education* 1, no. 1: 65–86.

Solomon, G. T., and L. W. Fernald, Jr. 1991. "Trends in Small Business Management and Entrepreneurship Education in the United States." *Entrepreneurship: Theory and Practice* 15 (3): 25–39.

Souitaris, V., S. Zerbinati, and A. Al-Laham. 2007. "Do entrepreneurship programs raise entrepreneurial intention of science and engineering students? The effect of learning, inspiration and resources." *Journal of Business Venturing* 22 (4): 566–91.

Thursby, M., A. Fuller, and J. Thursby. 2009. "An Integrated Approach to Educating Professionals for Careers in Innovation." *Academy of Management Learning and Education* 8 (3): 389–405.

Todorovic, Z. W., and N. Suntornpithug. 2008. "The multi-dimensional nature of university incubators: Capability/resource emphasis phases." *Journal of Enterprising Culture* 16 (4): 385–410.

Tracey, P., and N. Phillips. 2007. "The distinctive challenge of educating social entrepreneurs: A postscript and rejoinder to the special Issue on entrepreneurship education." *Academy of Management Learning and Education* 6(2): 264–71.

Tsai, W. 2002. "Social Structure of 'Coopetition' within a Multiunit Organization: Coordination, Competition, and Intra-Organizational Knowledge Sharing." *Organization Science* 13: 179–91.

Tushman, M. L., and P. Anderson. 1986. "Technological discontinuities and organizational environments." *Administrative Science Quarterly* 31 (3): 439–65.

Tushman, M. L., and C. A. O'Reilly, III. 2004. "The Ambidextrous Organization." *Harvard Business Review* 82 (4) (April): 74–81.

Verzat, C., J. Byrne, and A. Fayolle. 2009. "Tangling with Spaghetti: Pedagogical Lessons from Games." *Academy of Management Learning and Education* 8 (3): 356–69.

Vesper, K. H., and W. B. Gartner. 1997. "Measuring progress in entrepreneurship education." *Journal of Business Venturing* 12 (5): 403–21.

Vesper, K. H., and W. E. McMullan. 1988. "Entrepreneurship: Today Courses, Tomorrow Degrees?" *Entrepreneurship: Theory and Practice* 13 (1): 7–13.

Wadhwa, V., R. Freeman, and B. Rissing. 2008. *Education and Tech Entrepreneurship*. Kansas City, MO: Ewing Marion Kauffman Foundation.

Wee, K., and N. Lynda. 2004. "A problem-based learning approach in entrepreneurship education: Promoting authentic entrepreneurial learning." *International Journal of Technology Management* 28 (7/8): 685–701.

Weick, K. E. 1976. "Educational Organizations as Loosely Coupled Systems." *Administrative Science Quarterly* 21 (1): 1–19.

Wennekers, S., and R. Thurik. 1999. "Linking entrepreneurship and economic growth." *Small Business Economics* 13 (1): 27–56.

Westhead, P. 1997. "R & D 'Inputs' and 'Outputs' of Technology-Based Firms Located On and Off Science Parks." *R&D Management* 27: 45–62.

Westhead, P., and D. J. Storey. 1994. *An Assessment of Firms Located On and Off Science Parks in the United Kingdom*. London: HMSO.

Westhead, P., D. J. Storey, and M. Cowling. 1995. An Exploratory Analysis of the Factors Associated with the Survival of Independent High-Technology Firms in Great Britain. In *Small Firms: Partnerships for Growth*, edited by F. Chittenden, M. Robertson, and I. Marshall, 63–99. London: Paul Chapman.

World Economic Forum. 2009. *Educating the Next Wave of Entrepreneurs: Unlocking Entrepreneurial Capabilities to Meet the Global Challenges of the 21st Century*. A report of the Global Education Initiative, April. Geneva: World Economic Forum.

Wright, M., S. Birley, and S. Mosey. 2004. "Entrepreneurship and university technology transfer." *Journal of Technology Transfer* 29 (3–4): 235–46.

Wright, M., A. Vohora, and A. Lockett. 2004. "The formation of high-tech university spinouts: The role of joint ventures and venture capital investors." *Journal of Technology Transfer* 29 (3–4): 287–310.

Wright, M., et al. 2006. "University spin-out companies and venture capital." *Research Policy* 35(4): 481–501.

Wright, M., et al. 2007. "The role of human capital in technological entrepreneurship." *Entrepreneurship Theory and Practice* 31 (6): 791–805.

CHAPTER SIX

Research, Science, and Technology Parks

Vehicles for Technology Transfer

Albert N. Link and John T. Scott

1. Introduction

Research, science, and technology parks are increasingly seen as a means to create dynamic clusters that accelerate economic growth and international competitiveness through the transfer of knowledge and technology.[1] As such, it is important to understand the academic literature related to research, science, and technology parks (hereafter R-S-T parks, or simply parks) because that literature, albeit embryonic, has influenced and will continue to frame public policies related to park formations and growth.

The purpose of this chapter is thus to overview the extant academic literature on knowledge and technology transfer to and from parks, and to discuss its importance to public policy issues.[2] The remainder of this chapter is outlined as follows. In section 2, dimensions of a definition of a park are set forth. In section 3, alternative theories on R-S-T park formations are overviewed. Section 4 summarizes the extant empirical literature and places it in the context of a model of innovation. Section 5 offers examples of technology transfer from a university into its R-S-T park. Finally, this chapter concludes with policy considerations in section 6.

2. Definitions

The term *research park* is more prevalent in the United States, the term *science park* is more prevalent in Europe, and the term *technology park* is more prevalent in Asia. Many definitions of a park have been proffered, mostly by professional organizations (e.g., AURRP 1998; IASP 2005; UKSPA 2006; and UNESCO 2006) and by parks themselves as a way to define their activities. Common among these definitions is that a park is a type of public-private partnership that fosters knowledge flows—often between park firms and universities and among park firms—and contributes to regional economic growth and development.

Link and Scott (2006), based on an overview of alternative definitions of university research parks, and most parks in the United States are affiliated with a university, proposed the following definition:

> A university research park is a cluster of technology-based organizations that locate on or near a university campus in order to benefit from the university's knowledge base and ongoing research. The university not only transfers knowledge but expects to develop knowledge more effectively given the association with the tenants in the research park.

A public-private partnership, with reference to a park, is an infrastructure that leverages, formally or informally, the efficiency of innovation that takes place within park firms and, when present, within universities (Link and Link 2009). The term *public* refers to any aspect of the innovation process that involves the use of governmental resources, be they federal or national, state, or local in origin. The term *private* refers to any aspect of the innovation process that involves the use of private-sector resources, mostly firm-specific resources. And, resources are broadly defined to include all resources—financial resources, infrastructural resources, research resources, and the like—that affect the general environments in which innovation occurs. Finally, the term *partnership* refers to any and all innovation-related relationships, including but not limited to formal and informal collaborations in R&D.

In the case of parks in the United States, government involvement tends to be indirect with economic objectives of leveraging public-sector R&D (including university R&D) and private-sector R&D. In many

Asian countries, for example, government involvement is direct rather than indirect.[3]

3. Theories of R-S-T Park Formations

Surprisingly, the extant literature in economics, geography, entrepreneurship, management, and public policy does not offer a fully developed theory on the formation of parks. Case studies have documented the institutional history of a number of research parks, university affiliated or not. Castells and Hall (1994) and Wonglimpiyarat (2010) describe the Silicon Valley (California) and/or Route 128 (around Boston, Massachusetts) phenomenon; Luger and Goldstein (1991), Link (1995; 2002), and Link and Scott (2003a) detail the history of Research Triangle Park (North Carolina); Gibb (1985), Grayson (1993), Guy (1996a; 1996b), and Vedovello (1997) summarize aspects of the science park phenomenon in the United Kingdom; Gibb (1985) also chronicles the science/technology park phenomenon in Germany, Italy, the Netherlands, and selected Asian countries; and Chordà (1996) reports on French science parks; Phillimore (1999) on Australian science parks; Sternberg (2004) on technology centers in Germany; Bakouros, Mardas, and Varsakelis (2002) and Sofouli and Vonortas (2007) on the development of science parks in Greece; Hu (2007) on technology parks in China; Vaidyanathan (2008) on technology parks in India; Salvador (2011) on science parks in Italy; and Alshumaimri, Aldridge, and Audretsch (2010) on parks in Saudi Arabia.

Scholars have not yet formally tied the emergence of parks to cluster theory, although cluster theory has been applied to the formation of biotechnology and other science-based agglomerations of firms near universities, so the application is reasonable. Drawing on cluster theory—and location theory was, in part, a prequel to the popularization of cluster theory, as reviewed by Goldstein and Luger (1992) and Westhead and Batstone (1998)—one could argue that there are both demand and supply forces at work that result in the clustering of research firms near universities (Baptista 1998).

On the demand side, there are sophisticated users of developed technologies within a park, and the search costs for such users are minimized by locating on a park. Of course, there are disadvantages associated with being in a park, mainly greater competition for and with the developed technologies. When a park attracts many firms which then have access to the same technologies, those firms may expect greater competition in the use of those technologies. On the supply side, there is skilled and specialized labor available from the university or universities involved in the park in the form of graduate students and consulting faculty, although there is also more competition for that pool of human capital. Also, for a firm, location on a park, especially a university park, provides a greater opportunity for the acquisition of new knowledge—tacit knowledge, in particular. For the university, having juxtaposed firms provides a localized opportunity for licensing university-based inventions or even innovations. The theory of agglomeration economics emphasizes knowledge spillovers and enhanced benefits and lowered costs caused by the presence of multiple organizations and the externalities they create (Swann 1998). Empirical support for the agglomeration effects is provided by Audretsch (1998); Audretsch and Feldman (1996; 1999); Breschi and Lissoin (2001); Jaffe (1989); Jaffe, Trajtenberg, and Henderson (1993); and Rothaermel and Thursby (2005a; 2005b).

Henderson (1986) and Krugman (1991) emphasize conceptually as well as empirically the importance of location per se with regard to knowledge spillovers. Localization has an effect on resource prices. To the extent that new technology embodies new knowledge, geographic closeness implies lower new technology prices and thus presumably greater usage. Supporting the hypothesis that location matters, Link and Scott (2006) find that the employment in a park grows more rapidly the closer geographically the park is to the affiliated university. Moreover, Link and Scott (2005) find that university spin-off companies are more prevalent in a park as geographic proximity of the park and the university increases. Firms achieve economies of scale more easily with newer technologies. Arthur (1989) underscores the related importance of network externalities with regard to such scale economies. David (1985) also argues in general—and his argument could apply particularly well to university parks—that chance or historical events can lock a technology on a particular path of development. If that technology had a university origin, then creating such a park, from the university's perspective, and locating in the park, from a firm's perspective, gives positive feedback to continue the path dependency of the particular technology.[4]

Relatedly, Leyden, Link, and Siegel (2008) outline a theoretical model, based on the theory of clubs, to describe the conditions under which a firm would be located in an existing university park. The authors conjecture that a university park acts like a private organization, so that membership in the research park is the result of mutual agreement between the existing park tenants including the university, the club, and a potential new member firm. The decision to admit a new firm depends on the marginal effect of that firm on the well-being of the firms already in the park. For the representative in-park firm, the value of belonging to the park is the opportunity to engage in synergistic activities, which can be used to increase its profits in the output markets in which it participates, net of the direct cost (e.g., maintenance cost of being in the park and maintaining infrastructure) and indirect cost (e.g., congestion and competition for and with new knowledge) of being in the park.

Layson, Leyden, and Neufeld (2008) extend the park membership issue. Their theoretical model considers the optimal size of a research park in which each firm maximizes the net benefits it receives from being in the park and the park as a whole maximizes the average net benefits of all of the park tenants. Their model concludes that although there are benefits from knowledge and technology transfer from a host university to park tenants, the optimal size of the park is not related to whether or not a university is present (i.e., the optimal number of tenants is not necessarily larger or smaller).

Often absent from the theoretical discussions in the literature about park formations has been a focus of market failure. That is, public support of parks might be viewed from an economic perspective in terms of removing barriers that inhibit the two-way flow of scientific and technical knowledge between university faculty and tenant scientists, as well as among tenant scientists in the park. Martin and Scott (2000: 439) use the work of several scholars to support the idea that when innovation occurs through technology with high-science content, innovation market failure stems from the fact that the knowledge underlying the technology originates outside the commercial sector. Given the use of knowledge from outside the commercial sector, the creators of the knowledge may not recognize potential applications or effectively communicate the knowledge to potential users. Martin and Scott identify biotechnology, chemistry, materials science, and pharmaceuticals as typical sectors with application of high-science content technology. They observe that to address the innovation market failure associated with such technology with high-science content, the appropriate policy instrument is high-tech bridging institutions such as research parks that can facilitate the diffusion of advances in high-science-content research.

More generally, Link and Scott (2011: 8–12) list and discuss in detail several factors that create barriers to innovation and new technology, and the firms emerging from R-S-T parks face many of these barriers. For example, their projects often face high technical risk, long times to finish the R&D and commercialize the resulting technology, spillovers of the benefits of their technology to others, and the risk of opportunistic behavior when sharing information about their technology. The result is the expectation that without the support of public-private partnerships embodied in the incubators of R-S-T parks, the entrepreneurial firm would expect that its private rate of return will fall below its hurdle rate, even though the social rate of return exceeds the social hurdle rate. Of course, whether or not such conditions of innovation market failure are being addressed by R-S-T parks is a matter for careful evaluation, and we discuss such evaluation in the concluding section.

4. Empirical Studies of R-S-T Parks

The empirical literature related to parks is somewhat embryonic. Table 6.1 summarizes findings from the extant literature in four dimensions:

- Factors affecting firm decisions to locate on a park
- Formation of university parks and university performance
- Firm performance on a park
- Parks and regional economic growth development

It is clear from the literature review in table 6.1 that R-S-T parks, especially university parks, matter in several dimensions related to innovative activity and to economic growth and development.

To place the importance of these empirical findings in a broader perspective, consider the model of economic development in figure 6.1.[5] At the root of the model is the science base, referring to the accumulation of scientific knowledge. The science base resides in the public domain. Investment in the science base comes through basic research, primarily funded by the government and primarily performed globally in universities and federal laboratories.

For an integrated technology-based manufacturing firm—and the model is similar for a service sector firm (Gallaher, Link, and Petrusa 2006) except that technology is generally purchased rather than induced

TABLE 6.1. **Empirical studies of research, science, and technology parks.**

Research Issue	Author(s)	Dimensions of Study	Findings
Factors affecting firm decisions to locate on a park	Westhead and Batstone (1998)	Comparison of U.K. on-park and off-park firms	Location on a park is driven by the firm's desire to acquire research facilities and scientists at the university—all U.K. parks are located on or near a university
	Goldstein and Luger (1992)	Comparison of university-based and non-university based parks in the United States	Key criterion for location on a park is the linkage between the firm and the university (or, if generalizable to other countries, the higher education institution)
	Hansson, Husted, and Vestergaard (2005)	Case studies of U.K. and Denmark parks	Firms locate in the park because of a need for social capital to facilitate entrepreneurial growth
	Leyden, Link, and Siegel (2008)	U.S. public firms that have and do not have a research facility on a university park	Parks invite firms to join a park based on their potential spillover benefits (i.e., knowledge spillover benefits) to existing park firms
Formation of university parks and university performance	Link and Scott (2003b)	Growth of U.S. university parks over time	Growth of park formations follows a Gompertz survival-time model; formal park-university relationships lead to increased university publication and patenting activity, greater extramural funding success, improved placement of doctoral graduates, and enhanced ability to hire preeminent scholars
	Link and Scott (2005)	Formation of U.S. university spin-off companies within the university's research park	University spin-off companies are a greater proportion of a park's companies in parks that are older, are associated with better university research environments, have a biotechnology focus, and are geographically closer to their university,
Firm performance on a park	Westhead (1995), Westhead and Cowling (1995), Westhead and Storey (1994, 1997), and Westhead, Storey, and Cowling (1995)	Matched pair comparison of on-park and off-park U.K. firm performance	Survival rate of on-park firms greater than of off-park firms
	Siegel, Westhead, and Wright (2003)	Matched pair comparison of on-park and off-park U.K. firm performance	Research productivity of on-park firms greater than of off-park firms

Research Issue	Author(s)	Dimensions of Study	Findings
	Lindelöf and Löfsten (2003, 2004)	Matched pair comparison of on-park and off-park Swedish firms	On-park firms place greater emphasis on innovative ability, sales and employment growth, market orientation, and profitability than off-park firms
	Ferguson and Olofsson (2004)	Matched pair comparison of on-park and off-park Swedish firms	No performance differences between on-park and off-park firms
	Fukugawa (2006)	Matched pair comparision of on-park and off-park Japanese firms	Research linkages more likely formed with universities if on a park than off of a park
	Squicciarini (2008)	Matched pair comparision of on-park and off-park Finnish firms	Patenting activity in on-park firms is greater than in off-park firms
	Yang, Motohashi, and Chen (2009)	Matched pair comparison of on-park and off-park firms at or near the Hsinchu Science Industrial Park in Taiwan	Patenting per R&D dollar is greater in on-park firms
Parks and regional economic growth and development	Goldstein and Luger (1992) Shearmur and Doloreux (2000)	Descriptive analysis of U.S. park directors Descriptive analysis of Canadian park directors	Parks leverage new business start-ups Parks leverage new business start-ups and overall employment growth
	Appold (2004)	Empirical analysis of industrial research laboratories in the United States	New industrial laboratories are more likely to locate in a county with a university research park.
	Link and Scott (2006)	Growth of employment of U.S. university parks	Parks closer to a university grow faster

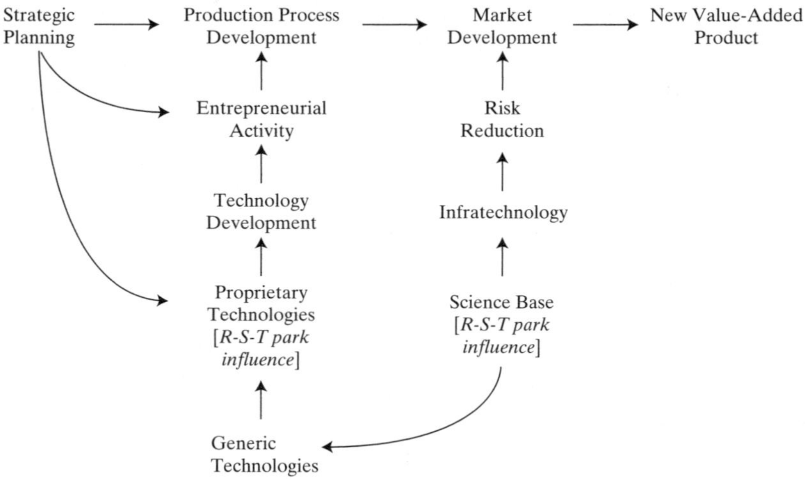

FIGURE 6.1. Model of Innovation: A Manufacturing Firm.

through own R&D—technology development in the form of basic and applied research generally begins within its laboratory. There, R&D involves the application of scientific knowledge toward the proof of concept of a new technology. Such fundamental research, if successful, yields a prototype or generic technology. If the prototype technology has potential commercial value, follow-on applied research takes place followed by development. If successful, a proprietary technology will result. Basic research, applied research, and development occur within the firm as a result of its strategic planning and guide the firm's market-oriented entrepreneurial activities. Generally, strategic planning involves the formulation of road maps for developing new emerging technologies. A manufacturing firm targets discrete technology jumps, creating new technologies that make its competition obsolete; its strategic plans are long-term and not closely linked to current competitive planning.

Entrepreneurial activity then drives the firm toward the production of the new product or process. With entrepreneurial activity, the overall innovation process exhibits hysteresis because of the lagging impact of such entrepreneurship. Thus, the relationship between investment in proprietary technology and market development might not be as predictable as the firm's strategic planners would like.

Infratechnologies (i.e., infrastructural technologies) emanate from the science base and from various technology infrastructures such as national laboratories. These technologies, such as test methods or measurement standards, reduce the market risk associated with the introduction of a new product or process. Once a new product has been designed and tested, technical risk could be low, but market risk could be significant until the product is accepted and then adapted and integrated into existing systems (e.g., in a service sector firm). The nonlinearity of this system is, literally, in the fact that there are multiple influences on both innovation and technology development, thus underscoring the existence of a need for broad-based and multi-targeted public sector innovation and technology policy actions.

The impact of R-S-T parks within this model is on the science base and on proprietary technology. Parks—university parks, in particular—enrich the science base because they leverage the university's R&D through a two-way transfer of knowledge. Parks also leverage the firm's R&D, again through technology transfer, especially in comparison to off-park firms.

5. Examples of Technology Transfer into a R-S-T Park

The following are three examples of university-based technology being transferred into the university's R-S-T park. These are certainly not the only examples of spin-out technology transfer, but they illustrate clearly that parks are an important infrastructure to support technology transfer.

Avid Radiopharmaceuticals, Inc. spun out of the University of Pennsylvania in 2005 and moved into University City Science Center business incubator. University City Science Center, established in 1963, is the largest urban park in the United States. In 2009, Avid relocated from the incubator into the park. Avid develops molecular imaging agents that allow for the earlier detection of diseases including Alzheimer's disease, Parkinson's disease, Dementia with Lewy Bodies and Diabetes mellitus.[6]

Liquidia Technologies was founded in 2004 as a University of North Carolina at Chapel Hill spin-out into Research Triangle Park. Liquidia has established a powerful and versatile nanotechnology-based product development platform that revolutionized engineered health care products. By leveraging fabrication techniques from the semiconductor industry, Liquidia can rapidly design and manufacture precisely engineered particles of virtually any size, shape, or composition.[7]

Adimab was founded by a Dartmouth College professor who previously spun out a biotechnology company into the R-S-T park affiliated with Dartmouth College—the Centerra Resource Park. Adimab began in Centerra's incubator, the Dartmouth Regional Technology Center, and then it moved out of the incubator and into its own facilities, still within the park. Adimab's integrated antibody discovery and optimization platform provides unprecedented speed from antigen to purified, full-length human IgGs. Adimab offers fundamental advantages by delivering diverse panels of therapeutically relevant antibodies that meet the most aggressive standards for affinity, epitope coverage, species cross-reactivity, and expressability. Adimab enables its partners to rapidly expand their biologics pipelines through a broad spectrum of technology access arrangements.[8]

6. Policy Conclusions

The elements of a national innovation system include competitive firms and a competitive environment, an effective educational system, strong university research, a legal system with property rights, and a capital market that includes venture capital (Nelson 1993; Cohen 2002). R-S-T parks have a unique place within a national innovation system. In their discussion of the design of public support for private innovation, Martin and Scott (2000: 444–46) observe:

> Where innovation relies on a technology base with a high science content, there is . . . a need for bridging institutions. . . . The role of bridging institutions in this case is to facilitate diffusion of advances in basic research from academic research operations to the private sector. . . . Close formal and informal connections with university researchers (including the physical location of some company research operations near universities) [have been observed to be] important factors making these linkages possible. . . . Bridging institutions here could be university-industry research parks. . . . Where innovation requires the application of knowledge and techniques on the frontier of knowledge, it is high-tech bridging institutions (university-industry commercial parks and the like) that are called for.

Although the literature related to parks is still embryonic, the evidence suggests that parks enhance the two-way transfer of knowledge between firms and between firms and universities, when a university is present. Thus, parks enhance innovation and subsequently competitiveness as was suggested in figure 6.1.

Many nations' sectors have to varying degrees informally encouraged the formation of industry/university linkages. France's central government, like that of Japan, the Netherlands, and the United Kingdom, has actively fostered the creation of science parks (Westhead 1997; Hilpert and Ruffieux 1991; Goldstein and Luger 1990) and Germany has long promoted academic innovation centers to incubate and develop small- and medium-sized enterprises (Sternberg 1990).

In the United States, public investment at state universities is used to underwrite the formation and development of R-S-T parks. In 2004, through Senator Bingaman's introduction of S. 2737, "The Science Park Administration Act of 2004," and again in 2007 through Senator Pryor's introduction of S. 1373, "The Building a Stronger America Act," the U. S. Congress considered, but did not pass, a bill to provide grants and loans to states and local authorities for the development and construction of university parks. Implicit in these bills is the assumption that R-S-T parks are an important element in the U. S. national innovation system, and as such should be fostered because of both the knowledge-based and employment-based spillovers that will result.

This U. S. action may the most obvious example of direct public-sector support for R-S-T parks. It may also be the most obvious example of indirect support of university-industry technology transfer since the passage of the Bayh-Dole Act in 1980. Hand in hand with public-sector support is the need for public accountability, namely the development and implementation of evaluation methods and tools not only to support the assumption that R-S-T parks are in fact an important element of the national innovation system but also to quantify the net spillover benefits that result from public-sector support.

The matched pairs studies discussed above in table 6.1 are a preliminary form of evaluation. That is, it is useful to know that there is evidence that firms located on a park are more productive than firms not on a research park, *ceteris paribus*. However, when substantial public-sector resources are devoted to park formations, a more in-depth evaluation approach is warranted, namely the application of what Link and Scott (2001; 2011) call the spillover evaluation method. The spillover evaluation method applies to publicly funded, privately performed research projects, and research projects are defined in terms of the research activities that occur in the park rather than simply the construction of the park.[9]

There are important projects where economic performance can be improved with public funding of privately performed research. Public funding is needed when socially valuable projects would not be undertaken without it. If their expected rate of return from creating an R-S-T park environment falls short of their required rate, called the hurdle rate, then the university or local firms would not invest in the research park environment. Nonetheless, if the benefits of the research spill over to consumers and to firms other than the ones investing in the research, the social rate of return may exceed the appropriate hurdle rate, even though the private rate of return falls short of the private hurdle rate. It would then be socially valuable to have the investments made, but since the university or local firms will not make them without public support, the public sector should support the investments. By providing public funding, thereby reducing the investment needed from the university and local firms doing the research, the expected private rate of return can be increased above the hurdle rate. In this case, the public sector's support may also suggest, or affirm, the possibility of a market for a successful project, thus reducing the investors' perceived risk as well as increasing the initial investment they are willing to make. Thus, because of the public subsidy, the university, when present, and local firms are willing to perform the research that is socially desirable because much of its output spills over to other firms in the park and sectors in the local and national economy.

The question asked in the spillover evaluation method is one that facilitates an economic understanding of the potential returns to public-sector support for a portion of private-sector research, namely: What is the proportion of the total profit stream generated by the university's and local firms' research and innovation that the university and local firms expect to capture; and hence, what proportion is not appropriated but is instead captured by others that use knowledge generated by park research to produce competing products for the social good?[10]

We conclude that R-S-T parks should not a priori be considered a primary element of a nation's innovation system, but rather that point of view, which evidently is held by a significant group in the U.S. Congress as well as by policy makers in other nations, needs more study. Successful two-way knowledge flow between universities and industry is a key ingredient for a national innovation system, and we do have evidence that R-S-T parks play a role in that knowledge flow. However, parks are not a sine qua non of the knowledge flow. Perhaps, consistent with the findings of the survey of university provosts reported in Link and Scott (2003b), R-S-T parks fall under the broader category of an effective educational system. However, R-S-T parks may in the future warrant a higher status, especially as technological life cycles continue to shorten and as basic research at universities—and to a growing extent at national laboratories (Wessner 1999; 2001)—and applied research/development in industry become more intertwined.

Note

1. This theme motivated the 2009 National Research Council symposium on research, science, and technology parks (Wessner 2009).

2. This chapter draws from Link and Scott (2007).

3. Direct government involvement in park activity is illustrated through the many summaries of activities in Asian parks in this chapter.

4. The idea of path dependency, according to Arrow (2000), has its origins in the early writings of economists Veblen and Cournot, but it also can be traced to the Nelson and Winter (1982) concepts about evolutionary economics.

5. For a more complete discussion of this model see Tassey (2007).

6. University City Science Center press release, August 13, 2009.

7. http://www.liquidia.com/Overview.html.

8. http://www.adimab.com/.

9. If one defined narrowly the output of the use of public-sector resources as the park itself, then, following Link and Scott (1998; 2011), the counterfactual evaluation method would be appropriate. When publicly funded, publicly performed research investments are evaluated, and the public is building the park, one should ask: What would the private sector have had to invest to achieve the benefits associated with the park in the absence of the public sector's investments? The answer to this question gives the benefits of the public's investments, namely, the costs avoided by the private sector.

10. The part of the stream of expected profits captured by the innovator is its private return, while the entire stream is the lower bound on the social rate of return (because of the additional benefits of consumer surplus and assuming any cannibalization of existing surplus is relatively small). The spillovers evaluation weighs the private return (in practice—see Link and Scott 2001—estimated through extensive interviews with the private sector organizations receiving public support regarding their expectations of future patterns of events and future abilities to appropriate returns from R&D-based knowledge) against private investments. The social rate of return weighs the social returns against the social investments.

References

Alshumaimri, A., T. Aldridge, and D. B. Audretsch. 2010. "The University Technology Transfer Revolution in Saudi Arabia." *Journal of Technology Transfer* 35 (6): 585–96.

Appold, S. J. 2004. "Research Parks and the Location of Industrial Research Laboratories: An Analysis of the Effectiveness of a Policy Intervention." *Research Policy* 33 (2): 225–43.

Arrow, K. J. 2000. "Increasing Returns: Historiographic Issues and Path Dependence." *European Journal of the History of Economic Thought* 7 (2): 171–80.

Arthur, W. 1989. "Competing Technologies, Increasing Returns, and Lock-in by Historical Small Events." *Economic Journal* 99 (2): 116–31.

Association of University Related Research Parks (AURRP). 1998. "Worldwide Research and Science Park Directory 1998." BPI Communications report.

Audretsch, D. B. 1998. "Agglomeration and the Location of Innovative Activity." *Oxford Review of Economic Policy* 14 (2): 18–29.

Audretsch, D. B., and M. P. Feldman. 1996. "R&D Spillovers and the Geography of Innovation and Production." *American Economic Review* 86 (3): 630–40.

Audretsch, D. B., and M. P. Feldman. 1999. "Innovation in Cities: Science-Based Diversity, Specialization, and Localized Competition." *European Economic Review* 43 (2): 409–29.

Bakouros, Y. L., D. C. Mardas, and N. C. Varsakelis. 2002. "Science Parks, a High Tech Fantasy? An Analysis of the Science Parks of Greece." *Technovation* 22 (2): 123–28.

Baptista, R. 1998. "Clusters, Innovation, and Growth: A Survey of the Literature." In *The Dynamics of Industrial Clustering*, edited by G. M. P. Swann, M. Prevezer, and D. Stout. Oxford: Oxford University Press.

Breschi, S., and F. Lissoin. 2001. "Knowledge Spillovers and Local Innovation Systems: A Critical Survey." *Industrial and Corporate Change* 10 (4): 975–1005.

Castells, M., and P. Hall. 1994. *Technopoles of the World*. London: Oxford University Press.

Chordà, I. M. 1996. "Towards the Maturity State: An Insight into the Performance of French Technopoles." *Technovation* 16 (3): 143–52.

Cohen, W. 2002. "Thoughts and Questions on Science Parks." Presented at the National Science Foundation Science Parks Indicators Workshop, University of North Carolina at Greensboro.

David, P. A. 1985. "Clio and the Economics of QWERTY." *American Economic Review* 75 (2): 332–37.

Ferguson, R., and C. Olofsson. 2004. "Science Parks and the Development of NTBFs: Location, Survival and Growth." *Journal of Technology Transfer* 29 (1): 5–17.

Fukugawa, N. 2006. "Science Parks in Japan and Their Value-Added Contributions to New Technology-Based Firms." *International Journal of Industrial Organization* 24 (2): 381–400.

Gallaher, M. P., A. N. Link, and J. E. Petrusa. 2006. *Innovation in the U.S. Service Sector.* London: Routledge.

Gibb, M. J. 1985. *Science Parks and Innovation Centres: Their Economic and Social Impact.* Amsterdam: Elsevier.

Goldstein, H. A., and M. I. Luger. 1990. "Science/Technology Parks and Regional Development Theory." *Economic Development Quarterly* 4 (1): 64–78.

Goldstein, H. A., and M. I. Luger. 1992. "University-Based Research Parks as a Rural Development Strategy." *Policy Studies Journal* 20 (2): 249–63.

Grayson, L. 1993. *Science Parks: An Experiment in High Technology Transfer.* London: British Library Board.

Guy, I. 1996a. "A Look at Aston Science Park." *Technovation* 16 (5): 217–18.

Guy, I. 1996b. "New Ventures on an Ancient Campus." *Technovation* 16(6): 269–70.

Hackett, S. M., and D. M. Dilts. 2004. "A Systematic Review of Business Incubation Research." *Journal of Technology Transfer* 29 (1): 55–82.

Hall, B. H., A. N. Link, and J. T. Scott. 2001. "Barriers Inhibiting Industry from Partnering with Universities: Evidence from the Advanced Technology Program." *Journal of Technology Transfer* 26 (1–2): 87–98.

Hall, B. H., A. N. Link, and J. T. Scott. 2003. "Universities as Research Partners." *Review of Economics and Statistics* 85 (2): 485–91.

Hansson, F., K. Husted, and J. Vestergaard. 2005. "Second Generation Science Parks: From Structural Holes Jockeys to Social Capital Catalysts of the Knowledge Society." *Technovation* 25 (9): 1039–49.

Henderson, J. V. 1986. "The Efficiency of Resource Usage and City Size." *Journal of Urban Economics* 19 (1): 47–70.

Hilpert, U., and B. Ruffieux. 1991. "Innovation, Politics and Regional Development: Technology Parks and Regional Participation in High-Technology in France and West Germany." In *Regional Innovation and Decentralization: High Technology Industry and Government Policy*, edited by U. Hilpert. London: Routledge.

Hu, A. G. 2007. "Technology Parks and Regional Economic Growth in China." *Research Policy* 36 (1): 76–87.

International Association of Science Parks (IASP). 2005. http://www.iasp.ws/publico/index.jsp?enl=2.

Jaffe, A. B. 1989. "Real Effects of Academic Research." *American Economic Review* 79 (5): 957–70.

Jaffe, A. B, M. Trajtenberg, and R. Henderson. 1993. "Geographic Localization of Knowledge Spillovers as Evidenced by Patent Citations." *Quarterly Journal of Economics* 108 (3): 577–98.

Krugman, P. 1991. *Geography and Trade*. Cambridge, MA: MIT Press.

Layson, S. K., D. P. Leyden, and J. Neufeld. 2008. "To Admit or Not to Admit: The Question of Research Park Size." *Economics of Innovation and New Technology* 17 (7–8): 689–97.

Leyden, D. P., A. N. Link, and D. S. Siegel. 2008. "A Theoretical and Empirical Analysis of the Decision to Locate on a University Research Park." *IEEE Transactions on Engineering Management* 55 (1): 23–28.

Lindelöf, P., and H. Löfsten. 2003. "Science Park Location and New Technology-Based Firms in Sweden: Implications for Strategy and Performance." *Small Business Economics* 20 (3): 245–58.

Lindelöf, P., and H. Löfsten. 2004. "Proximity as a Resource Base for Competitive Advantage: University-Industry Links for Technology Transfer." *Journal of Technology Transfer* 29 (3–4): 311–26.

Link, A. N. 1981a. "Basic Research and Productivity Increase in Manufacturing: Some Additional Evidence." *American Economic Review* 71 (5): 1111–12.

Link, A. N. 1981b. *Research and Development Activity in U.S. Manufacturing*. New York: Praeger.

Link, A. N. 1995. *A Generosity of Spirit: The Early History of the Research Triangle Park*. Research Triangle Park, NC: University of North Carolina Press for the Research Triangle Park Foundation.

Link, A. N. 2002. *From Seed to Harvest: The History of the Growth of the Research Triangle Park*. Research Triangle Park, NC: University of North Carolina Press for the Research Triangle Park Foundation.

Link, A. N., and J. R. Link. 2009. *Government as Entrepreneur*. New York: Oxford University Press.

Link, A. N., and K. R. Link. 2003. "On the Growth of U.S. Science Parks." *Journal of Technology Transfer* 28 (1): 81–85.

Link, A. N., and J. T. Scott. 1998. *Public Accountability: Evaluating Technology-Based Institutions*. Norwell, MA: Kluwer.

Link, A. N., and J. T. Scott. 2001. "Public/Private Partnerships: Stimulating Competition in a Dynamic Market." *International Journal of Industrial Organization* 19 (5): 763–94.

Link, A. N., and J. T. Scott. 2003a. "The Growth of Research Triangle Park." *Small Business Economics* 20 (2): 167–75.

Link, A. N., and J. T. Scott. 2003b. "U.S. Science Parks: The Diffusion of an Innovation and Its Effects on the Academic Mission of Universities." *International Journal of Industrial Organization* 21 (9): 1323–56.

Link, A. N., and J. T. Scott. 2005. "Opening the Ivory Tower's Door: An Analysis of the Determinants of the Formation of U. S. University Spin-Off Companies." *Research Policy* 34 (7): 1106–12.

Link, A. N., and J. T. Scott. 2006. "U. S. University Research Parks." *Journal of Productivity Analysis* 25 (1): 43–55.

Link, A. N., and J. T. Scott. 2007. "The Economics of University Research Parks." *Oxford Review of Economic Policy* 23 (4): 661–74.

Link, A. N., and J. T. Scott. 2011. *Public Goods, Public Gains: Calculating the Social Benefits of Public R&D*. New York: Oxford University Press.

Link, A. N., and D. S. Siegel. 2003. *Technological Change and Economic Performance*. London: Routledge.

Luger, M. I., and H. A. Goldstein. 1991. *Technology in the Garden*. Chapel Hill: University of North Carolina Press.

Martin, S., and J. T. Scott. 2000. "The Nature of Innovation Market Failure and the Design of Public Support for Private Innovation." *Research Policy* 29 (4–5): 437–47.

Nelson, R. R. 1993. *National Innovation Systems: A Comparative Analysis*. New York: Oxford University Press.

Nelson, R. R., and S. G. Winter. 1982. *An Evolutionary Theory of Economic Change*. Cambridge, MA: Harvard University Press.

Phillimore, J. 1999. "Beyond the Linear View of Innovation in Science Park Evaluation: An Analysis of Western Australian Technology Park." *Technovation* 19 (11): 673–80.

Rothaermel, F. T., and M. C. Thursby. 2005a. "Incubator Firm Failure or Graduation? The Role of University Linkages." *Research Policy* 34 (7): 1076–90.

Rothaermel, F. T., and M. C. Thursby. 2005b. "University-Incubator Firm Knowledge Flows: Assessing Their Impact on Incubator Firm Performance." *Research Policy* 34 (3): 302–20.

Salvador, E. 2011. "Are Science Parks and Incubators Good 'Brand Names' for Spin-Offs? The Case of Turin." *Journal of Technology Transfer* 36 (2): 203–32.

Shearmur, R., and D. Doloreux. 2000. "Science Parks: Actors or Reactors? Canadian Science Parks in their Urban Context." *Environment and Planning* 32 (6): 1065–82.

Siegel, D. S., P. Westhead, and M. Wright. 2003. "Assessing the Impact of Science Parks on Research Productivity: Exploratory Firm-Level Evidence from the United Kingdom." *International Journal of Industrial Organization* 21 (9): 1357–69.

Sofouli, E., and N. S. Vonortas. 2007. "S&T Parks and Business Incubators in Middle-Sized Countries: The Case of Greece." *Journal of Technology Transfer* 32 (5): 525–44.

Squicciarini, M. 2008. "Science Parks' Tenants Versus Out-Of-Park Firms: Who Innovates More? A Duration Model." *Journal of Technology Transfer* 33 (1): 45–71.

Sternberg, R. 1990. "The Impact of Innovation Centers on Small Technology-Based Firms: The Example of the Federal Republic of Germany." *Small Business Economics* 2 (2): 105–18.

Sternberg, R. 2004. "Technology Centers in Germany: Economic Justification, Effectiveness and Impact on High-Tech Regions." *International Journal of Technology Management* 28 (7): 444–69.

Swann, G. M. P. 1998. "Towards a Model of Clustering in High-Technology Industries." In *The Dynamics of Industrial Clustering*, edited by G. M. P. Swann, M. Prevezer, and D. Stout. Oxford: Oxford University Press.

Tassey, G. 2007. *The Technology Imperative*. Northampton, MA: Edward Elgar.

United Nations Educational, Scientific and Cultural Organization (UNESCO). 2006. http://www.unesco.org/science/psd/thm_innov/unispar/sc_parks/concept shtml/.

United Kingdom Science Park Association's (UKSPA). 2006. http://www.ukspa .org.uk/about_ukspa/faqs_about_ukspa/.

Vaidyanathan, G. 2008. "Technology Parks in a Developing Country: The Case of India." *Journal of Technology Transfer* 33 (3): 285–99.

Vedovello, C. 1997. "Science Parks and University-Industry Interaction: Geographical Proximity between the Agents as a Driving Force." *Technovation* 17 (9): 491–502.

Wessner, C. 1999. *A Review of the Sandia Science and Technology Park Initiative*. Washington, D. C.: National Academy Press.

Wessner, C. 2001. *A Review of the New Initiatives at the NASA Ames Research Center: Summary of a Workshop*. Washington, D. C.: National Academy Press.

Wessner, C. 2009. *Understanding Research, Science, and Technology Parks: Global Best Practices*. Washington, D. C.: National Academy Press.

Westhead, P. 1995. "New Owner-Managed Businesses in Rural and Urban Areas in Great Britian: A Matched Pairs Comparison." *Regional Studies* 29 (4): 367–80.

Westhead, P. 1997. "R&D 'Inputs' and 'Outputs' of Technology-Based firms Located On and Off Science Parks." *R&D Management* 27 (1): 45–61.

Westhead, P., and Batstone, S. 1998. "Independent Technology-Based Firms: The Perceived Benefits of a Science Park Location." *Urban Studies* 35 (12): 2197–219.

Westhead, P., and M. Cowling. 1995. "Employment Change in Independent Owner-Managed High-Technology Firms in Great Britain." *Small Business Economics* 7 (2): 111–40.

Westhead, P., and D. Storey. 1994. *An Assessment of Firms Located On and Off Science Parks in the United Kingdom*. London: Her Majesty's Stationery Office (HMSO).

Westhead, P., and D. Storey. 1997. "Financial Constraints on the Growth of High-Technology Small Firms in the U. K." *Applied Financial Economics* 7 (2): 197–201.

Westhead, P., D. J. Storey, and M. Cowling. 1995. "An Exploratory Analysis of the Factors Associated with the Survival of Independent High-Technology Firms in Great Britain." In *Small Firms: Partnerships for Growth*, edited by F. Chittenden, M. Robertson, and I. Marshall. London: Paul Chapman.

Wonglimpiyarat, J. 2010. "Commercialization Strategies of Technology: Lessons from Silicon Valley." *Journal of Technology Transfer* 35 (2): 225–36.

Yang, C. H., K. Motohashi, and J. R. Chen. 2009. "Are New Technology-Based Firms Located on Science Parks Really More Innovative? Evidence from Taiwan." *Research Policy* 38 (1): 77–85.

CHAPTER SEVEN

University Patenting in Europe

Does Faculty Ownership of Intellectual Property Impede University Technology Transfer?

David Audretsch and Devrim Göktepe-Hultén

Dr. Göktepe-Hultén gratefully acknowledges financial support from VINNOYA, the Swedish Governmental Agency for Innovation Systems.

1. Introduction

Over the last 30 years, academic and policy interest in the commercialization of new technologies from universities in the U.S. has increased considerably (Siegel et al. 2003a). An important reason for this rise in university technology transfer was the passage of the Bayh–Dole Act of 1980, which decreased the uncertainty associated with the commercialization of federally funded research. This legislation encouraged universities to be more proactive in their efforts to commercialize scientific discoveries (Mowery et al. 2004) by transferring ownership from the government to universities and other contractors who could then license the intellectual property (IP) to firms.

Although the effects of the Bayh-Dole Act on the increase of patenting are far from definite and conclusive, universities and other public research organizations (PROs) are increasingly protecting their inventions—from genetic discoveries to software programs—with the expectation of generating additional funds for research as well as the formation of new ventures. For instance, prior to 1981, fewer than 250 patents were issued to universities each year. In contrast, slightly over a decade later, almost 1,600 patents were being issued each year (Henderson, Jaffe, and Trajtenberg 1998; Shane 2004). Between 1993 and 2000, U.S. universities were granted some 20,000 patents. Over that period, some of these university patents generated millions of dollars in licensing revenues and spurred the creation of over 3,000 new firms, according to AUTM data (OECD 2003). The AUTM Licensing Survey 2002 reports that 569 new commercial products were launched that year, 450 new companies were established (for a total of 4,320 since the introduction of Bayh-Dole Act in 1980, of which 2,741 were still operating in 2002), running royalties on product sales were $1.005 billion, and new licenses and options executed in 2002 increased 15.2% over 2001. As a result, many observers have concluded that there may be a positive relation between the number of university patents and the Bayh-Dole Act. Yet there have been skeptical views that the Bayh-Dole Act is not the only factor behind the rise of university patenting in the U.S. and many other factors also came into play in the upsurge of patenting and licensing in the post-1980 period (Mowery et al. 2004).

On the European side, it has been argued that the level of commercial activity at European universities is relatively low compared to the high levels of scientific performance and investment in research, known as the "European Academic Paradox." In large part, this awareness reflected the recognition by governments that, in some cases, placing the outputs of publicly funded research in the public domain is not sufficient to generate social and economic benefits from research (OECD 2003). This perception is somewhat exacerbated by the impression that universities in the U.S. have performed much better in commercializing their research results because of the Bayh-Dole Act in contrast to individual ownership of IP at the European Universities. Subsequently since early 2000, many European countries (e.g., Austria, Denmark, Germany, and Norway) have abolished the so-called professor's privilege that granted academics the right to own patents. The right to ownership of IP has now been transferred to the universities while academic inventors are given a share of royalty revenue in exchange. There has also been an ongoing debate in Sweden on whether to follow a similar path and transfer ownership to universities, yet no decision has been made hitherto, and Sweden is the only Nordic country that still keeps the professor's privilege.

Such reforms are actually not only confined to European countries. For example, Japan has made legislative reforms to allow universities to protect and claim IP. China has amended Science and Technology Law of 1993 to the Bayh-Dole Act as of July 1, 2008. It is believed to be giving greater incentives to researchers to boost innovation. India has also introduced "The Protection and Utilization of Publicly Funded Intellectual Property Bill, 2008" (Sampat 2009). On March 23, 2010, the Philippines enacted an act providing the framework and support system for the ownership, management, use, and commercialization of IP generated from R&D funded by the government and for other purposes. However, the implementation of such reforms remains a challenge.

Some scholars criticized these reforms and pursued a different methodology to identify university-invented patents rather than relying on university-owned patents (Meyer, Sinilainen, and Utecht 2003; Meyer et al. 2005). They argued that in the organizational ownership of the IP model (Bayh-Dole Act), the university or its TTOs own almost all of the patented inventions made by its faculty. However, it is more difficult to find the university connection to the patented inventions in the individual ownership of IP regimes. They found that although university-owned patents are limited, university-invented patents are quite substantial. These studies came to the conclusion that the European Academic Paradox might not necessarily exist as the differences in the amount of university patents can be explained by the differences in legislations. Despite the importance of the Bayh-Dole Act in fostering technology transfer in the U.S., different national innovation systems may require different solutions. Abolishment of the professor's privilege is, therefore, arguably unnecessary and futile.

Based on these studies, in this chapter we investigate what other implications the professor's privilege may have besides causing differences in the number of patents owned or invented by universities. In section 2, we present the patent legislations and recent changes at the universities and public research organizations in some European countries. In section 3, we first address to what extent the law of professor's privilege has fostered university inventions to be pre-emptied by incumbent firms and decreased the likelihood and intentions of entrepreneurship among academics? In line with these questions we then discuss whether this law causes tendency of "solution oriented research" at the universities, and only contributes to the in-house research activities of existing firms, rather than science-driven research. We finally relate our discussion to whether the critical view on TTOs is grounded on the argument that incumbent companies' monopoly on the direct appropriation of university knowledge will be challenged.

In the concluding section, we show that the nature of university patenting depends not only on the ownership regimes but also on the informal and formal collaborative patterns of industrial relations. We discuss the strengths and weaknesses of the current system, and present some implications for university administrators and policy makers.

2. Ownership of Patents at European Universities and Legislative Changes

Unlike the Bayh-Dole Act of the U.S., many European countries have or used to have dual intellectual property rights (IPR) systems. While ownership of IP in the nonuniversity sector (i.e., firms and public research organizations) belongs to the organization (employer), university researchers can retain ownership rights to patents. Inspired partly by the U.S. legislation and partly as a result of the converging policies of the E.U. and OECD, a number of countries have passed legislation similar to the Bayh-Dole Act, and several other countries are considering or discussing similar changes with the expectation of more technology transfer from universities, which may generate financial benefits. In this section we give an overview of recent legislative changes among different European countries. This overview can be completed with the summary table 7A.1.

In Belgium, universities fall under competence of community governments. In Flanders all intellectual property from university researchers belongs to university. Since 1998 universities in the Walloon region can own the results of research that is fully funded by the region. The 1999 Decree on Education was adopted to create a framework for intellectual property at universities. A special decree was introduced regarding IP arising from research at public labs granting them to title and requiring them to agree upfront on the ownership of IP in collaboration research with universities.

Denmark introduced the Act on Inventions at PROs (universities, hospitals) effective as of January 1, 2000. The new law grants PROs to own employee inventions (Valentin and Jensen 2007). The law concerning inventions at public research of June 1999 focuses on the increasing cooperation between research institutions and businesses to make new knowledge and competence available to Danish society. Among other things, the law permits research institutions to take over rights to inventions made by their employees, and obliges the institutions to try to commercialize the inventions they have taken over.

The federal German government launched a knowledge transfer initiative, "Knowledge Creates Markets," in 2001. One of the actions of this initiative was the reform of a section of German employer-employee law (*Arbeitnehmererfindungsgesetz*) dealing with inventions by teaching faculty at universities (the so-called professor's privilege). In Germany, in 2002, a decision was made to change the ownership of IP within higher education institutes (HEI) by removing the exclusive ownership rights of researchers and transferring those rights to the employing organizations, though researchers retain rights to receive two-thirds of any licensing or other income from their invention (OECD 2003). With this reform the universities got ownership of the IP generated by their academics. A federal program starting 2002 was set to promote the commercialization of university research through the creation of Patent Marketing Agencies. Concerning cooperation projects with industry, the IP ownership depends on the agreement decided by the university and the industry at the beginning of the project.

Austria also introduced university ownership of IP in 2002. Accordingly inventions made at a university in the course of federal employment or training by such university shall be subject to the Patents Act, and the university shall be deemed to be the employer according to the Patents Act. That means that universities can claim rights for service inventions without the need for a separate written agreement between the university and university member. The main technology transfer office of Austria is held by Austria Wirtschaftservice Gesellschaft mbH (aws) with its department, Tecma, owned by the Republic of Austria. Tecma is in charge of patent exploitation and supports and accompanies researchers, inventors, and companies in marketing promising innovations. Tecma also assisted the Austrian university to install local departments. Their responsibility is to help academic inventors and to support technology transfer.

In Norway, a new bill on the commercial exploitation of inventions went into effect in January 2003. Under certain conditions, it transfers the right to commercialize an invention from researchers to the employing organization. In doing so, it has sought to establish organizational ownership by universities of the intellectual property from the results of research carried out at universities. Even if the new legislation adheres formally to patentable research results, PROs and HEIs are advised to deal with all commercializable results that cannot even be patented. A division of income accruing from commercialization was suggested with one-third going to the individual scientist(s), the department, and the institution. Organizational set-ups like TTOs were also being established at the Norwegian universities based on government funding.

In the Netherlands, the technology transfer activities have been improved by the Innovation Charter, which was signed in November 2004 by the universities, the employers' association and the federation of university medical centers. In the charter, the three parties put forward criteria for cooperation between industry and knowledge institutions and recommendations for arrangements concerning intellectual property, scientific integrity, and publication strategy. The universities adopt various rules for patent expenditure and income. The law states that inventions made through a university belong to the university. Patent applications have become the responsibility of the universities, as is their funding. A number of universities have consequently set up a patent fund, which is sometimes maintained through the proceeds from earlier patents.

Finland has introduced university ownership with effect from 2007 as well. A distinction is made between open research and collaborative research. The universities are entitled to acquire the rights to the invention that is made in collaborative research within six months from the disclosure of the invention. When the university has acquired rights in the invention, the inventor will be entitled to obtain reasonable compensation for the invention. The university also has the competence to make contractual arrangements with third parties concerning patentable inventions that may be made within research projects. The entitlement to inventions resulting from open research will remain with the inventor. Such inventions, however, are covered by the duty to inform the HEI about the invention.

In France, the legislation is quite general (i.e., universities and public research organizations are considered employers, which will own the rights on inventions made by staff). Revenues resulting from patent licensing are shared in equal parts between the university, the department that sponsored the research, and the team of researchers and professors who produced the invention. Furthermore, individual scientists are encouraged to patent their inventions, while the university itself promotes all the efforts to pursue an effective transfer of patented technologies. In the case of CNRS, patents are either licensed to external firms, or even sold in return for funding (Cesaroni and Piccaluga 2005).

Similar to the U.S. model, in the U.K., universities generally own the IP generated by their academics. In 1948, the British government set up the National Research Development Corporation (NRDC) to commercialize publicly funded research in Britain. After merging with the National Enterprise Board in 1975, this organization was renamed British Technology Group International which functions as a technology transfer organization. It was privatized in 1992. However, until the 1980s they were obliged to use the services of the NRDC to patent and license technologies. Until the late 1990s, technology transfer activities were funded entirely by the home university, though the government (Department of Trade and Industry) advised universities to play a greater direct role in exploiting research. Most technology transfer units were quite small and underfunded. Nowadays universities are able to undertake the full range of technology commercialization activities in their own name and most have their own TTOs. Different universities structure their TTOs in different ways—some as divisions of the university, others as wholly owned subsidiaries. One or two have even gone semiprivate, selling a part of themselves to a third-party investor.

Sweden has kept the law of the professor's privilege allowing university researchers to retain the IPR to their research results. At the same time, Sweden has also created many new TTOs, including university holding companies and other regional technology transfer agents. However, this mandate has been expressed in more general terms that refer to the general interaction with and communication of research results to the broader society (Jacob, Lundqvist, and Hellsmark 2003).

Unlike many other European countries, in 2001, Italy first introduced the individual ownership of IP at the universities on the basis of the opposite intuition that individual scientists may have a greater incentive to patent than the universities that employ them. However, it also encouraged the formation of technology transfer infrastructure at the universities. In 2005, the rules have changed again, and the title of all the inventions coming from externally funded research and developed within a public institute are given to the organization, leaving the researcher the possibility of a title only for research funded by the intra-institution source.

These policy measures are motivated not only by the arguments that these new institutions and organizations can support and speed up the industrial exploitation of academic research, but also that the financial returns from patenting may help to support research and teaching at universities. Mowery et al. (2004) have highlighted that the passage of the Bayh-Dole Act coincided with a number of developments in academic research, industry and policy, as well as a decentralized system for funding of universities, administrative autonomy, and the need for external resources that facilitate university patenting.

As for the outcomes of individual ownership of patents or new legislations at European universities, there is not reliable data on the amount of university patents or their distribution by ownership. Based on these critical aspects, a number of scholars pursued a different methodology to identify the actual number of university patents at the European universities through identifying the inventors. In the next section we give an overview of these empirical studies.

3. University Inventors and Patenting in the E.U.

As mentioned in the previous sections in Europe, levels of university patenting, licensing, and spin-off company formation have been claimed to be low compared to the relatively high level of investment in higher education institutes or in basic research. This phenomenon has been labeled as European Academic Paradox, according to which European countries have a strong science base but are not good at transferring research results into commercially viable new technologies (Tijssen and van Wijk 1999). Although there has been no systematic attempt at measurement until recently, it is well known that no European university holds a patent portfolio as large as MIT's or Stanford's. It is also believed that many European universities do not hold any patents at all (OECD 2003; Lissoni et al. 2007). Given the impression of higher numbers of university patents and spin-offs, and higher licensing revenues for the U.S. universities, an emulation of the U.S. Bayh-Dole Act has been advised by many European policy makers.

However, some concerns have been raised, that such policy suggestions and recent changes (summarized in section 2) are based to a large extent on unrealistic and faulty assumptions that are based on inadequate or erroneous information (Geuna and Nesta 2006; Verspagen 2006). Most information on university patents comes from surveys submitted to university administration, newly established TTOs, or on cursory searches for the names of universities or university TTOs as the applicants for patents. Owing to the different institutional and organizational setups at European universities, university patenting should be investigated by finding the names of university scientists who are also registered as inventors in patent databases instead of searching for university names or newly established university TTOs as applicants of patents. A series of European studies on university patenting have been conducted to estimate the amount of university patenting in Europe and construct patent data sets comparable with those of U.S. universities (Meyer, Sinilainen, and Utecht 2003; Balconi, Breschi, and Lissoni 2004; Meyer et al. 2005; Iversen, Gulbrandsen, and Klitkou 2007; Lissoni et al. 2007; Göktepe 2008; Lissoni et al. 2009).

These studies distinguished between patents owned by universities and patents that were invented by university researchers but not owned by the university. University-owned patents are the patents in which universities or university TTOs are listed as applicants of these patents. Such patents are usually applied for and managed by a TTO. University-invented patents are conversely defined through the affiliation of their inventors with a university. Such patents have at least one university employee as an inventor. Depending on either individual or organizational ownership of IP, scientists can apply for patents individually, university investors may assign their rights to another party to apply for a patent with the aid of TTOs, or scientists can apply through other actors such as firms.

The empirical strategy has relied on a unique way of combining the names and addresses of inventors in the registers of national patent offices, European Patent Office (EPO) and/or United States Patents and Trademark Organization (USPTO) with the names and addresses of university researchers' registers. The results of name and address matching between university researchers and inventors were then validated for each patent by direct contacts with the inventors to confirm whether the matching between inventors and university employees is correct. After controlling for the homonyms (e.g., two inventors with the same name and surnames that may match an academic researcher/professor with that name and surname) by surveys, e-mails, and telephone, the identity of the matches are confirmed (see Meyer, Sinilainen, and Utecht 2003 and Göktepe 2008 for a detailed description of the methodology of identification of university patents). As a result of these methodological exercises, new and unique university-patent data sets were constructed. However, these data sets have some limitations. Some of them fail to capture all patents invented by academic scientists who retired or moved out of the university system, and thus, were not found in the university records; or some of them fail to confirm if a university researcher is an inventor when in fact two people have the same names. Moreover, owing to different time periods or patent databases, a direct cross-country comparison will be limited. Nevertheless, these university data sets provide reliable and useful information on the extent of patenting among the scientists vis-à-vis TTOs, distribution of applicants/ownership, and technological classification.

The common finding of these studies is that there are more university-invented patents than university-owned patents across European universities. The inventive output of European universities or university researchers is higher than previously thought. Before these studies, university patents were often understood as patents assigned to universities, and the patenting activities of university researchers were more or less invisible in European studies (see Cesaroni and Piccaluga 2002; Saragossi and von Pottelsberghe de la Potterie 2003). In what follows, we briefly present the findings of each individual study. (This overview can be completed with the summary table 7A.2.)

In a number of studies, Meyer and his collaborators used a matching procedure between first and family names of inventors in patent databases and university researcher registers. They matched all USPTO patents that had at least one Finnish inventor for the period 1986 to 2000 with the names of university researchers that were employed at Finnish universities in the years 1997 and 2000. First, Meyer, Sinilainen, and Utecht (2003) reported that Finnish universities owned 36 USPTO patents that had at least one Finnish inventor, while university-invented patents amounted to 530. Second, in their comparative study of Flemish and Finnish universities, Meyer et al. (2005) found that there were 379 university-invented patents compared to 100 university-owned patents at Flemish universities. Balconi, Breschi, and Lissoni (2004) found that out of 1,475 university-invented patents in Italy between 1978 and 1999, only 40 EPO patents had universities as applicants, whereas Italian university-inventor patents account for 3.8% of EPO patents by Italian inventors.

Azagra-Caro, Carayol, and Llerena (2006) pointed out that although French universities are legally entitled to own patents based on scientists' research results, the university-invented, but not university-owned, patent has been and remains in practice the most common form of patenting at the University Louis Pasteur (ULP) in France. ULP had 463 patents from the French Patent Office, the EPO and other patent offices from 1993 to 2000. Of these, only 62 patents were owned by the ULP.

Giuri et al. (2007) showed that the total number of university patents in the PatVal survey of inventors for six European countries (Italy, United Kingdom, the Netherlands, France, Germany, and Spain) was 433. The PatVal survey was addressed to inventors listed on (granted) European patents with a priority date in the period of 1993–1997 in six European countries—Germany, France, Italy, the Netherlands, Spain, and the U.K. These six countries accounted for about 88% of granted EPO patents whose first inventor has an address in one of the E.U.-15 countries (about 42% of the total EPO). The survey obtained responses relating to 9,017 patents representing 18% of all granted EPO patents with a priority date in the considered period. Out of 9,017 patents, 433 patents, which were identified as university patents, have at least one inventor who was employed by a university. Crespi, Geuna, and Nesta (2007) further investigated the 433 university patents and found that much of the university research that leads to patents in Europe does not show up in the statistics because it is private firms rather than the universities themselves that apply for the patent. About 80% of the EPO patents with at least one academic inventor are not owned by the university.

Iversen, Gulbrandsen, and Klitkou (2007: 405) found that a total of 569 researchers from Norwegian public research organizations were involved in at least one patent application in the years between 1998 and 2003. These researchers were involved in 10 to 11% of domestic patent applications during those years. The contribution of university and college researchers was high in chemical and pharmaceutical patenting, accounting for nearly 18%. Göktepe (2008) showed that a total of 458 patents were identified as owned by Lund University in Sweden, and 250 university researchers were identified as inventors in the EPO patent database (between 1990 and 2004). University patents account for about 2% of the total number of national patents between September 1990 and September 2004.

Based on the KEINS database, Lissoni et al. (2007) found that the university professors who were active in Sweden and Italy during 2004 and in France during 2005 were responsible for a substantial number of patent applications during the period between 1978 and 2002. During that period there were 2,800 patent applications in France; 2,200 in Italy; and 1,400 in Sweden. Lissoni et al. (2007) compared the level of patenting in these three countries (between 1994 and 2001) with the U.S.

university patent data (between 1993 and 2000) in order to make a comparison possible between the U.S. and Europe. They found that French, Italian, and Swedish university-owned patents constituted less than 1% of the total number of domestic patents. The proportions of university-invented patents are around 3% in France, 4% in Italy, and more than 6% in Sweden. U.S. estimates for university-invented patents are about 6% (Thursby, Fuller, and Thursby 2009). Similar to Crespi, Geuna, and Nesta (2007), Lissoni et al. (2007) have also shown that the alleged gap between the U.S. and Europe in terms of university patenting turns out to be a very limited gap between the U.S., France, and Italy and no gap at all between the U.S. and Sweden.

In their follow-up study for Denmark, Lissoni et al. (2009) identified 306 Danish academic inventors among the 7,395 professors who were active in 2001, 2004, and 2005. They found 495 patents that have at least one professor as an inventor of the patent.

Despite the organizational ownership of IP at U.S. universities, this phenomenon has been also observed in the U.S., yet to a lesser extent. Aldridge and Audretsch (2010) also found that 70% of faculty chose to commercialize their research by assigning all patents to their university transfer offices (TTO) while 30% chose a "backdoor route" to commercialization and did not assign at least some of their patents to the university TTO. Scientists who did not assign patents to their university to commercialize research tend to rely on the commercialization mode of starting a new firm. Similarly Thursby, Fuller, and Thursby (2007) found that 26% of the patents are assigned solely to firms rather than to the faculty member's university as is dictated by U.S. university employment policies or the Bayh-Dole Act. Patents assigned to firms (whether established or start-ups with the inventor as principal) are less basic than those assigned to universities, suggesting that these patents resulted from faculty consulting.

Different technology transfer infrastructures and patent legislation at European universities require a different methodological approach. Finding the names of university inventors by matching them with the names of inventors in patent databases, rather than making cursory investigations of the names of universities or TTOs as applicants for patents, gives a better picture of the extent of university patenting in Europe. The lack of university patents in Europe is actually a lack of university-owned patents, not necessarily a lack of university-invented patents. Once the data are corrected to take into account the different ownership structure in Europe and the U.S., simple calculations suggest that the European academic system seems to perform much better than had been believed until now. In relative terms, European universities' patenting output lags only marginally behind that of U.S. universities (Crespi, Geuna, and Nesta 2007). These empirical investigations showed that *university patenting is not a new phenomenon for European universities.* They provide clear empirical evidence that the *number of university-invented patents is much higher than the number of patents owned by universities.* As such, scholars argued that European universities do not necessarily need to emulate the Bayh-Dole Act in order to increase university patenting. However, our understanding of university patenting in Europe is still incomplete for reasons that we now address.

3.1. Seamless Web between Scientists and Big Industry

First, we investigate the owners and applicants of the university-invented patents to address to what extent the professor's privilege (or the traditional European model of IP at universities) has fostered university patents to be pre-emptied by incumbent firms and also decreased the likelihood and intentions of entrepreneurship among academics. Inventors at most of the European universities may choose different routes to commercialize their research results. They can, for instance, patent individually, patent through TTOs, or patent through industrial firms.

According to KEINS database, in all of the three European countries considered, university administrations have much less control over professors' IPRs than in the U.S. In Sweden, where the professor's privilege is still standing, academic scientists often patent in their own name, as shown by the 14% share of patents assigned to individual scientists. Sixty percent of academic patents in France are owned by business companies. Seventy-four percent of Italian and 82% of Swedish academic patents are owned by firms; in contrast, business companies own only 24% of U.S. academic patents. Conversely, universities in three European countries own a very small share of academic patents: around 8% in France and Italy and less than 4% in Sweden, well below the 69% share in the hands of U.S. universities (Lissoni et al. 2007). Similarly, almost 70% of Danish university patents at the EPO are owned by firms. This is followed by patents filed by their own inventors (12%) and universities (12%). Governmental agencies account for less than 3% of university patents.

According to KEINS database, both in Italy and France, large state-controlled companies (such as ST-Microelectronics, ENI, France Telecom, and Tales) hold a very large number of academic patents. Large multinational companies (Ericsson and ABB) are also the important owners of the Swedish academic patents. Lissoni et al. (2009) found that the largest amount of academic patents in Denmark is owned by the most important pharmaceutical company NovaNordisk (including Novazymes). In all of these countries, only one university is among the top patent holders (i.e., the country's largest universities: Rome-La Sapienza in Italy; Paris 6 in France; Karolinska Institute in Sweden; and Copenhagen University in Denmark).

In line with the above findings, firms are the main owners of Lund University patents that were applied for during 1990 and 2004 vis-à-vis Lund University or TTOs. In all, 117 firms applied for 363 patents. This constituted 79.3% of all patents. Inventors applied for 12.9% of all patents, or 59 out of 458 patents. These patents were not assigned to any company at the time of application. Inventors most likely transfer (license, sell, or give) their patents to firms later. In total, 11 different TTOs located within the university, the region, or outside of Sweden applied for 36 patents out of 458. TTO patents constituted 7.9% of all university patents. According to the law of professor's privilege, any IP emanating from university-industry collaboration will be transferred automatically to partner companies in return for further research funding and collaboration. Yet these results are not recorded as invention disclosures to universities or patent applications by the universities.

At Lund University, large firms accounted for 57.6% of the patent applications; they applied for 209 of the 363 patents for which applications were submitted by firms. Twenty-seven SMEs applied for 51 patents, while 58 spin-offs applied for 103 patents. These spin-off firms were started by either former or current academics, and they maintained close relations with the department or research group they spun off from. Some spin-off firms have, for instance, employed students from the research group. Although the absolute number of university spin-offs (58) is almost double the number of large established firms (32), patents applied for by large companies is twice as that of patents owned by spin-offs. There are 105 patents in ICT and telecom-related fields. Of these ICT- and telecom-related patents, 80 were applied for by Sweden's largest telecom company (Ericsson). As in the case of ICT patents, most of the patents in pharmaceuticals were also applied for by large firms. In sectors such as mechanics, materials, and biotechnology, there were more SMEs and spin-off firms as patent applicants. In biotechnology and related fields, 39 patents were applied for by the 19 spin-off firms, and 14 SMEs applied for 24 patents. In materials and related fields, 17 patents were applied for by four spin-off firms.

Ericsson, ABB, Astra-Zeneca, and Gambro AB are the key applicants of the Lund University related patents. A few start-ups—Obducat AB, Amersham AB, Bioinvent AB—are the applicants of the Lund University patents. These figures reflect the dominance of large firms in the Swedish economy and their relatively easier access to university knowledge. Universities and scientists often try to keep good relations with the companies in order to have a steady flow of research contracts, funds, and materials, as well as to have job options for their students. Given these conditions, industry typically supports research at the universities, pays for the cost of the patent application and maintenance costs, and owns the patent, yet scientists behind the novel idea are recorded as the inventors. However, these firms do not have the obligation to develop and utilize the patent, and generally they do not pay royalty fees back to the university or scientists except so that they may further collaborate with the scientists behind the idea and, thus, sponsor research at the universities.

As industry funds more university research than in the past, defining the IPR provisions for cooperative research agreements or industrial research contracts with companies becomes important. Many public research funding agencies, universities, and public research labs have standard contracts for this. Contract law is especially important in defining the terms for licensing of IPR from or to universities. This is a frequent practice in the case of industry funding. For instance, the Swedish private sector is greatly dominated by large firms, and industrial R&D is also highly concentrated to a few very large firms; in 1994, four multinationals carried out more than 70% of total R&D among multinationals (Braunerhjelm 1998). The findings here indicate that most of the university technology transfer activities have been informal and are taken in by large firms without universities claiming any rights over these patents.

University-invented patents can also be analyzed by looking at the distribution across science and technology fields. Thursby, Fuller, and Thursby (2009) find that biotech patents from U.S. universities are more likely to be held by universities than electronic patents are, which in turn have a higher probability to be held by business companies. According to the KEINS database this is also the case for three European countries (France, Italy, and Sweden), in the four most academic-intensive technologies. Business companies own almost 80% of academic patents in electronics and electrical engineering but only a little more than 58% of those in pharmaceuticals and biotechnology, where both universities and government hold record shares of 14% and 20%, respectively. It is worth noting that academic patents in instruments also see a lower-than-average share of business ownership and the record share of individual ownership (over 9%).

The strongest technological sectors in each country also tend to be those in which university patents are heavily concentrated. For instance, patents in telecommunication in Finland account for 12% of university-invented patents while pharmaceuticals and biotechnology account for about 9% each (Meyer, Sinilainen, and Utecht 2003). The broadly defined research area of biotechnology and pharmaceuticals tends to be an area of extremely high university patenting activity in many countries. These studies have found almost the same tendencies as in the U.S. In 1998, 41% of U.S. academic USPTO patents were in three areas of biomedicine, indicating a strong focus on developments in the life sciences and biotechnology. In terms of revenues, about half of the total royalties were related to life sciences, including biotechnology (NSF 2002). Whether a corresponding degree of concentration in this area exists for university patents in Europe is less clear-cut, but the available evidence is not at odds with this assumption (Geuna and Nesta 2006). The technological specialization of the national systems of innovation and the dominance of a few large companies in these specific sectors partly explains why a substantial amount of university patents has been concentrated in a small number of large firms.

Like other European countries in Germany, university-owned patents are found to be relatively rare, but university-invented patents have been increasing continuously from less than 200 in the early 1970s to around 1,800 in 2000 (Meyer-Krahmer and Schmoch 1998). Czarnitzki, Ebersberger, and Fier (2007) and Czarnitzki, Glänzel, and Hussinger (2009) found a relatively low number of German university patents. Grimpe and Hussinger (2008) also argued that university licensing, which is receiving much attention in the literature on industry-science links, is used by relatively few German firms. Instead, consulting and informal collaboration have been shown to be substantial. Authors therefore assumed that university scientists in Germany, and especially highly credentialed faculty members, should be more likely to engage in informal technology transfer than their U.S. counterparts, Although several years have passed since the abolishment of the professor's privilege in Germany, Grimpe and Fier (2010) assumed that the specific German orientation toward excellence in engineering plays a role for the informal technology transfer behavior. Engineering research is organized in large research groups with multimillion Euros research funding and close collaboration with industry. Moreover, engineers in Germany are typically members of the German engineers association ("VDI—Verein Deutscher Ingenieure"). Both bonds through their alma mater and the network provided by the engineering association should therefore facilitate informal technology transfer activities.

As in the case of Sweden, we related patenting to the field of scientific specialization. Of the total number of patents from the three faculties under consideration, 63% emerged from the Faculty of Engineering, 32% are related to the Medical School, and 5% originated from the natural and basic science fields. A basic explanation for the distribution among scientific fields could be that in certain fields (basic theoretical physics, geology), patenting is not the preferred route for the protection and utilization of research results. It has also been more common to patent in fields like biotechnology, chemistry, or engineering. The Faculty of Natural Sciences has 36 patents. The lower rates of patenting at the Faculty of Natural Sciences can be explained, in part, by the more basic research conducted there compared to the applied and industrially relevant nature of research at the Faculties of Engineering and the Medical School. The intensity of patenting activities differs not only among these three faculties, but the departments within the same faculty differ considerably in their patenting activities as well. Each university patent was allocated to a university department by the identification of the inventor's departmental affiliation.

Researchers at Medical Faculty are also involved in centers such as the Strategic Centre for Clinical Cancer, which is funded by the Swedish Foundation for Strategic Research (SSF) and the Lund Centre for Stem Cell Biology and Cell Therapy, which is also funded by SSF. The aim of such platforms is to bring together different research groups and resources to do translational or interdisciplinary research by integrating clinicians and researchers from university hospitals with researchers from the Faculties of Medicine, Natural Sciences, and Engineering. Such research platforms, together with industrial partners, may induce patenting activities among researchers. In the subsequent sections of this paper we reflect more on the role and availability of industrial funding, the participation of scientists in collaborative projects with industrial partners, and the Swedish National Competence Centers to explain the differences among departments within the same faculty.

Researchers from the Faculty of Engineering participate in various types of research centers and research programs supported by Swedish National Agency for VINNOVA, regional actors, and industrial partners. There are 77 patents at the Department of Electrical Measurements and Industrial Electrical Engineering and Automation. Approximately 70 researchers were employed there during 1999–2004, on average. The research at this department is also industrially relevant, and several researchers participate in industrial networks. The Department of Engineering Physics has 44 patents. Researchers at this department are also engaged in centers, such as the Nanometer Structure Consortium or the Centre for Combustion Science and Technology, or they interact with industry through individual networks, which may influence the patenting activities at the department.

The Centre for Mathematical Sciences is a joint department covering all areas of mathematical sciences at LTH. There are about 100 employees and 41 patents at Mathematical Sciences. Most of the research is carried out within applied mathematics, especially image analysis and computer vision. Researchers also participate in joint research projects on the automatic generation of metadata from images, a Swedish Foundation for Strategic Research (SSF) funded project on computer and biological vision, and so on. Such projects may also increase the possibility of patenting at the Centre for Mathematical Sciences. Building and Environmental Technology had 20 patents, and there were about 100 employees during 1999–2004. The research at the department covers a broad spectrum of fundamental and applied issues in building technology. Research activities are strongly supported by the forest industry sector.

The ownership patterns of university patents seem to depend on the disciplinary affiliation of the inventors (and therefore also on the technological contents of the patents). Owing to the Swedish industrial specialization, interactions and consultancy in the fields of medicine, chemistry, and engineering were quite common. Many large Swedish firms have had strong connections with leading universities in Sweden for a long time (Stankiewicz 1986; Etzkowitz et al. 2000). It is likely that most of the research results go through the grapevine, and have been taken in by existing companies. Like in many other European countries, large multinational corporations tend to dominate the nature and route of research activities at the Swedish universities both directly and indirectly.

Moreover, current Swedish policies on the formation of university-industry competence centers or projects with industrial partners have strengthened such relations. These scientists have established long-term relations mainly with the large Swedish industrial corporations (e.g., Ericsson, ABB, and Volvo). Much of the funding comes directly from these industrial partners in return for the ownership of any intellectual property resulting from the project. One of the inventors we interviewed from the Faculty of Engineering who is working closely with the automobile industry commented that

> he and his colleagues are not interested in being listed as inventors, as this does not give any kind of merit. Such scientists are mainly in the expectation of generating further research funds and create job opportunities within the partner firms. They do not want to destroy the delicate bonds they or their supervisors worked hard to solidify, severely undermining their ability to secure further benefits they are expecting from their industrial partners. Most of them also believed that it is quite unlikely that they can start-up a firm within their field.

Some of the academic inventors are also cooperating with a group of industrial firms through competence centers. The partner firms have the first right of refusal for the inventions that are generated within these competence centers. Research results are disclosed to the partner firms and interested partner firms may apply for the patent. As emphasized by the scientists who are involved in collaborative arrangements with firms, the law of professor's privilege facilitates the relations with industrial partners since scientists can easily transfer the ownership rights of their intellectual property to the partner firms without the intervention of TTOs. In line with the above discussion, the law of professor's privilege and close informal relations with the existing firms, as well as the contractual structure of competence centers, lead to a tendency for a reverse linear model of innovation where research questions are generated by industrial needs. Such research contributes to the in-house research activities of existing firms, unlike science-pushed research where new ideas that emerge lead to new firms.

3.2. Role of TTOs

In this section, the role of the TTOs is analyzed and explained. In relation to the discussions above on the professor's privilege, it is important to understand the views of inventors on TTOs. Most successful inventors learned patenting and commercialization process individually or through their industrial contacts. It is very difficult for these academic inventors to accept the role and involvement of TTOs. They consider TTOs as impractical and unnecessary since the firms they collaborate with are the entities that apply for the patents. At the same time, it is unlikely that those firms will allow TTOs to sell the patents resulting from their collaborative projects. Although some inventors admitted that an efficient TTO can be helpful, they are skeptical about the organizational capacity of TTOs in the process of selecting inventions for patent application. The fear is that TTOs might be overloaded and might not respond to the needs of the scientists. In a situation like this, scientists, the university, and all parties would lose. By the same token, some inventors have implied that if a change happens in the ownership of patents—that is, if scientists no longer own their patents—scientists may avoid invention disclosures; they may bypass the TTO or disclose fewer and second-rate ideas to the TTO.

Furthermore, some inventors and their research groups have initiated their own TTOs where they try to commercialize themselves. These internal TTOs have become substitutes for a broader and more general university TTO. Incentives provided by these organizations are less formalized, and patenting activities of inventors depend mostly on contacts made on an individual basis or based on their innate skills and experiences. Many inventors have rated their private or individual contacts as being superior to a university's centralized unit that handles technology transfer on behalf of every scientist.

Scientists in general are critical about the assumptions that TTOs and laws may motivate scientists to patent. Although they do admit that TTOs can enable and facilitate scientists' patenting, TTOs are not the main driving force in deciding either to patent or not. If this was the case, scientists would have disclosed as many inventions as possible to the TTOs, and TTOs would be under pressure to apply for as many patents as one can imagine. It is clear that any political push to encourage scientists to commercialize their research would not work but rather create opportunists instead of scientists. If the rules and regulations force the scientists to get involved in commercial activities, they will, for instance, try to patent and even waste resources on start-ups. There will be the rush toward entrepreneurship since they have to report the number of the patents and commercial activities they have engaged in for the purposes of academic promotion. These patents will become "curriculum vita patents" or "mandated patents" unless they are utilized. As we elaborate further in section 4, these factors by and large both weakened the role TTOs can play, as well as created a critical view that TTOs are not facilitating technology transfer at the European universities.

4. Discussion and Concluding Remarks

In this chapter we aimed to investigate to what extent the so-called professor's privilege and other organizational factors may contribute to the pre-emption of the university inventions by the existing firms. We asked whether scientists prefer to bestow their IPR to industry partners, and how scientists' incentives and expectations influence their commercial practices. In the light of these findings, we argue that several factors impede new venture creation at universities.

The IPR regime at the universities, traditional and informal relations, establishment of competence centers, weaker structure of TTOs, and lack of entrepreneurial motivations among the scientists may cause the preemption of the university inventions by the incumbents and may impede new venture creation. These findings concerning the implications of the law of "professor's privilege" can be interpreted from two lenses:

(1) These findings are not necessarily negative. Research funding constraints are forcing many scientists to seek external collaboration. At the same time, many industrial firms rely more on collaborative research with universities instead of conducting basic research in their own laboratories or on asking universities to solve their problems. Accordingly, one can first argue that knowledge is codeveloped, and university research results transfer directly to the existing firms. Patents will be directly transferred to the industrial partner without unnecessarily waiting for the TTO or private patent attorney to evaluate and negotiate with the firms for licensing and other contracts. At the same time, one can expect close interaction with industry to lead more quickly to deeper scientific understanding and breakthroughs. Partner companies will further invest in product development and marketing which would

have been difficult with the university resources and scientists' own skills and interest in commercialization. Research groups and scientists often engage in competence centers or joint projects with several companies from within the same industry or complementary industries. Therefore, different aspects of knowledge and research results can be utilized by different firms rather than a single firm's exclusive appropriation.

Second, the steady relations between researchers and firms that were solidified over the years will continue. Moreover, scientists also benefit by signaling their knowledge and skills by showing their industrially relevant research and attract further funding. As no time will be lost for negotiations between universities and TTOs and the firms, one can also expect that in the professor's privilege system, quicker publication and dissemination of research may be possible. Concerning financial and other personal rewards, the professor's privilege may also bring some elasticity in the European university system, where salary structures and career paths can be quite rigid, by rewarding individual efforts and networks. The professor's privilege provides (or used to provide) different routes to university inventors like commercializing individually, through TTOs, or direct transfer to industrial firms.

(2) However the implications of the professor's privilege and close relations with existing firms may not be always positive. Formation of collaborative research, especially signed under the professor's privilege regime, may raise questions about a shift in focus from fundamental to applied research in universities. It is also possible that contracts and agreements in the U.S. may be based on more fundamental research than their equivalents in Europe, and thus generate broader patents. Broad patents may be more valuable to universities to the extent that they may be exploited through licensing for royalties rather than through cross-licensing for production purposes (Lissoni et al. 2007). However, in the case of joint projects at European universities, big established firms may decide the nature of the research and may lead to more narrower patents that are solving the existing industrial problems. Consequently, as large partner companies have stronger influence than smaller firms and scientists on the research areas and questions, they may obtain more than what they are actually supporting financially. Another concern would be that professors who are collaborating individually with a single firm can also exclusively transfer all the research results and IPR to a single company. Are too many talented researchers settling too easily into a routine of doing applied research for industry while ignoring fundamental issues that hold the keys to the next generation of new products (see above positive implication)?

Moreover, the prevalence of "professor's privilege and the formation of collaborative research with firms," while helping established companies to develop competence in their fields or in new fields, may decrease the possibilities available for academic entrepreneurs to utilize and start their firms. Similarly, it is likely that the most interesting, commercially viable research results are taken by established firms, while more mediocre or embryonic research results will be left to the university TTOs or research groups. As a result, there will be little that TTOs or scientists can do to develop and commercialize such research results. These are probably some of the reasons for the weakness and scarcity of high-technology start-ups from European universities. Furthermore, some researchers may not be interested in commercialization and may not have the necessary skills and contacts to do so, either. Moreover, as the costs of IP protection can be considerably high for an individual scientist or research group, many scientists and research groups may simply give up pursuing IP protection. Thus new commercial opportunities may be lost through these factors.

Finally, under the professor's privilege regime, one can expect more disputes over the identification of inventorship among the researchers, as well as with firms. It may severely undermine secure disclosures, and the knowledge flow can become severely blocked since licensees and investors simply do not work with IP that involves reluctant, distrustful, or downright uncooperative inventors.

Overall it is our impression that much is unknown about the advantages and disadvantages of the European system of university industry technology transfer. On one side, one group of scholars and policy makers supported the idea of adopting the Bayh-Dole Act as the solution to the "European Paradox." On the other side, another group of scholars criticized that such policy adoptions were made on the basis of the wrong presumption that European universities do not contribute enough to the production of patentable technology.

We, on the other hand, argued that the policy discussions should not solely be made on the basis of whether one system is producing patents or not. Although several studies found university-invented patents are much higher than the university-owned patents in the professor's privilege regime, implications of the professor's privilege and the close informal or formal collaborative relations with industry are beyond the quantifiable number of patents. The debate over why European universities should not retain the IPRs over their scientists' inventions and the weaker roles of TTOs can be partly explained by their dependence on their relationship with the industry as well as with their own academic staff.

Most European governments (as well as other countries such as Japan, India, and China) have pursued a portfolio of policies aimed at directing funds at university patenting specifically and entrepreneurial activities in general. The effectiveness of these policies can largely be influenced by the incentives of university researchers to commercialize their ideas. Without attention to the research life offered by universities and particularly the extent to which it attracts researchers to stay at universities, politicians will devise incomplete policy solutions. These policy tools may neglect and even threaten the main sources of satisfaction and personal development that most researchers derive from their research environments.

Consequently, while politicians and university administrators focus on legislation and organizational structures for academic entrepreneurship, the main component of entrepreneurship, research, has not received the attention and investment it needs. Today, research at universities receives less attention and investment than is given to, say, incubators and science parks. While science parks and incubators are very important nurturers of university knowledge, they need a constant infusing of new knowledge. The system works best when incentives are spread rather than there being attempts by one actor to steer academic scientists in a particular direction. In spite of the difficulties in evaluating the impact of policy instruments a priori, it is clear that these problems alone cannot justify the untenable practice of developing and maintaining one-size-fits-all policy instruments.

In order to provide better answers to the question of the management of IPR at the universities, one should look into the research activities from which the patents whose existence we have uncovered come from: Do they originate from research projects, whose results the universities prefer to leave in business partners' hands, possibly in exchange of a lump sum reward? Or do they originate from academic scientists' consultancy to business firms, which escape university administrations' control? How well/widely do these collaborative arrangements promote technology transfer? Are they economically viable for universities? What is the prevalence of collaborative research in various scientific fields? Does university-industry collaboration take only place in departments like biotechnology, nanotechnology, and electronics, while departments of basic sciences are marginalized? What are the participation ratios of large versus small companies in collaborative research projects (like in competence centers)? How is IP shared and utilized among the participants? What are the specific benefits of the professor's privilege to junior researchers and students?

The more we examine the European cases, the more we are forced to recognize that the observed patterns of university patenting in Europe depend much more on the institutional features of academic research and careers, traditional relations with industry, and even on the establishment of university-industry competence centers rather than on the success or failure of IPR reform and technology transfer policies.

Appendix

TABLE 7A.1. **Ownership of IP at European PROs and universities.**

Country	Current Legal Framework
Austria	University ownership since 2002
Belgium	University ownership since 1990s (both Flanders/Wallonia)
Bulgaria	No clear legal framework. Commercial activities are not part of the academic tasks.
Switzerland	Depends on employer-employee contracts and provision in agreements between funder of research and institute.
Czech Republic	No explicit university legislation—internal regulations within the universities.
Germany	University ownership since 2001
Denmark	University ownership with effect from 2001
Greece	No specific regulations appear to exist
Spain	Spanish Patent Law (1986) indicates that universities will apply for patents resulting from research of their professors
Finland	University ownership with effect from 2007.
France	Organizational ownership for PRO and university ownership
U.K.	Organizational ownership for PRO and university ownership
Hungary	University ownership, since 2006
Italy	Article 7 of National Law No 383 of Oct. 18, 2001, assigns title of inventions at universities to researchers. 2005 Organizational ownership.

Country	Current Legal Framework
Lithuania	No special legal framework for patenting at universities—internal regulations within the universities.
Latvia	No special legal framework for patenting at universities—internal regulations within the universities.
Netherlands	Legislation is general (i.e., universities are considered as employers, which will own the rights on inventions made by staff).
Norway	University ownership, since 2003.
Portugal	No clear legal situation.
Poland	No special legal framework for patenting at universities—internal regulations within the universities.
Sweden	Professor's privilege together with the formation of TTOs.
Slovakia	There is no special legal framework. Universities may claim ownership as employers of inventors.
Slovenia	University ownership, since 2007.

Adapted from OECD, 2003 Turning Science into Business

TABLE 7A.2. **Summary of Studies on University Patenting in Europe.**

Country	Time period database	#of university-invented patents	# of university-owned patents	Main technological category of patents	Type of university investigated
Finland (Meyer et a1. 2003a)	1986–2000 USPTO	530 patent 285 inventors	36	Telecom Instruments Pharmaceuticals	All universities except Social Science and Arts etc.
Flanders (Meyer et a1.2003b)	1986–2000 USPTO	379	100 (TTOs)	Organic chem. Life science	Technical universities
France (Azagra-Cara and Llerena 2003) (Univ.of Strasbourg)	1993–2000 French National Patent Office	463	62	Genetics Biology Physics	University of Strasbourg 82 Research laboratories
Germany (Schmoch 2000)	1970–2000 EPO	1800 (2000) and 200 (1970)	NA	Biotech, Medical engineering, Organic chemistry	All university professors. Title of professor is searched

(continued)

TABLE 7A.2. (*continued*)

Country	Time period database	#of university-invented patents	# of university-owned patents	Main technological category of patents	Type of university investigated
Italy (Balcani et al.) and other Boccani studies	1978–1999 EPO	1,475 919 inventors	40	Biotechnology Drugs, organic chemistry	All professors registered to Ministry of Education and Research
Norway (Gulbrandsen et a1. 2005)	1998–200 Norwegian domestic patents)	307 (8–12% of all Norwegian domestic patents)	NA	Life sciences Instruments	All researchers at universities/ colleges
Sweden-Chalmers (Wallmark-Survey)	1943–1994 Swedish patents or EPO	417 68	NA	Chemical engineering Electrical engineering	Chalmers University of Technology
Sweden-East Gothia (Schild) Linkoping University	1980–1996 (Swedish-PCT filings from East Gothia Region)	88 (Swedish-PCT filings from East Gothia Region) 82 inventor	NA	Instruments Electricity Health and amusement	Linköping University Technical Faculties

References

Aldridge, T., and D. B. Audretsch. 2010. "Does policy influence the commercialization route? Evidence from National Institutes of Health funded scientists." *Research Policy* 39 (5): 583–88.

Azagra-Caro, J. M., N. Carayol, and P. Llerena. 2006. "Patent protection at a European research university: Exploratory evidence at the laboratory level." *Journal of Technology Transfer* 31 (2): 257–68.

Balconi, M., S. Breschi, and F. Lissoni. 2004. "Networks of inventors and the role of academia: An exploration of Italian patent data." *Research Policy* 33 (1): 127–45.

Braunerhjelm, P. 1998. "Is there a relation between R&D-investments and high-technology exports? The Swedish export-paradox." *Ekonomiska Samfundets Tidskrift* 51 (2): 113–36.

Cesaroni, F. and A. Piccalugag. 2002. "Patenting activity of European universities: Relevant? Growing? Useful?" Paper presented in SPRU, NPRnet Conference, University of Sussex, March 21–23.

Cesaroni, F., and A. Piccaluga. 2005. "Universities and intellectual property rights in Southern European countries." *Technology Analysis and Strategic Management* 17 (4): 497–518.

Crespi, G. A., A. Geuna, and L. Nesta. 2007. "The mobility of university inventors in Europe." *Journal of Technology Transfer* 32: 195–215.

Czarnitzki, D., B. Ebersberger, and A. Fier. 2007. "The relationship between R&D collaboration, subsidies and R&D performance: Empirical evidence from Finland and Germany." *Journal of Applied Econometrics* 22 (7): 1347–66.

Czarnitzki, D., W. Glanzel, and K. Hussinger. 2009. "Heterogeneity of patenting activity and implications for scientific research." *Research Policy* 38 (1): 26–34.

Etzkowitz, H., et al. 2000. "The future of the university and the university of the future: Evolution of ivory tower to entrepreneurial paradigm." *Research Policy* 29 (2): 313–30.

Geuna, A., and L. Nesta. 2006. "University patenting and its effects on academic research: The emerging European evidence." *Research Policy* 35: 790–807.

Giuri, Paola, et al. 2007. "Inventors and invention processes in Europe: Results from the PatVal-EU survey." *Research Policy* 36 (8): 1107–27.

Göktepe, D. 2008. "Inside the ivory tower: Inventors and patents at Lund University." PhD diss. Lund, Sweden: Lund University.

Grimpe, C., and A. Fier. 2010. "Informal university technology transfer: A comparison between the United States and Germany." *Journal of Technology Transfer* 35 (6): 637–50.

Henderson, R., A. B. Jaffe, and M. Trajtenberg. 1998. "Universities as a source of commercial technology: A detailed analysis of university patenting 1965–1988." *Review of Economics and Statistics* 80 (1): 119–27.

Iversen, E. J., M. Gulbrandsen, and A. Klitkou. 2007. "A baseline for the impact of academic patenting legislation in Norway." *Scientometrics* 70 (2): 393–414.

Jacob, M., M. Lundqvist, and H. Hellsmark. 2003. "Entrepreneurial transformations in the Swedish university system: The case of Chalmers University of Technology." *Research Policy* 32 (9): 1555–68.

Lissoni, F., et al. 2007. "Academic patenting in Europe: New evidence from the KEINS database." Cespri Working Paper No. 202. Milan: Bocconi University.

Lissoni, F., et al. 2009. "Academic patenting in Europe: Evidence on France, Italy, and Sweden from the KEINS database." In *Learning to Compete in European Universities: Social Institution to Knowledge Business*, ed. M. McKelvey and M. Holmen, 187–218. Northampton, MA: Edward Elgar.

Meyer, M., T. Sinilainen, T., J. T. Utecht. 2003. "Towards hybrid Triple Helix indicators: A study of university-related patents and a survey of academic inventors." *Scientometrics* 58 (2): 321–50.

Meyer, M., et al. 2005. "Inventive output of academic research: A comparison of two science systems." *Scientometrics* 63 (1): 145–61.

Meyer-Krahmer, F., and U. Schmoch. 1998. "Science-based technologies: University–industry interactions in four fields." *Research Policy* 27 (8): 835–51.

Mowery, D. C., and B. N. Sampat. 2001. "Patenting and Licensing University Inventions: Lessons from the History of the Research Corporation." Oxford: Oxford University Press.

Mowery, D. C., et al. 2001. "The growth of patenting and licensing by US universities: An assessment of the effects of the Bayh-Dole act of 1980." *Research Policy* 30 (1): 99–119.

Mowery, D.C., et al. 2004. *Ivory Tower and Industrial Innovation: University-Industry Technology Transfer Before and After the Bayh-Dole Act*. Stanford, CA: Stanford University Press.

National Science Foundation (NSF). 2002. *Science and Engineering Indicators*. Arlington, VA: National Science Board.

Organisation for Economic Co-operation and Development (OECD). 2003. *Turning Science into Business. Patenting and Licensing at Public Research Organizations*. Paris: OECD Publications.

Sampat, B. N. 2009. "Academic Patents and Access to Medicines in Developing Countries." *American Journal of Public Health* 99 (1): 9–17.

Saragossi S., and B. van Pottelsberghe de la Potterie. 2003. "What Patent Data Reveal about Universities: The Case of Belgium." *Journal of Technology Transfer* 28 (1): 47–51.

Shane, S. 2004. "Encouraging university entrepreneurship? The effect of the Bayh-Dole Act on university patenting in the United States." *Journal of Business Venturing* 19 (1): 127–51.

Siegel, D. S., D. Waldman, and A. Link. 2003. "Assessing the impact of organizational practices on the relative productivity of university technology transfer offices: An exploratory study." *Research Policy* 32 (1): 27–48.

Siegel, D. S., et al. 2003a. "Commercial knowledge transfers from universities to firms: Improving the effectiveness of university-industry collaboration." *Journal of High Technology Management Research* 14 (1): 111–33.

Siegel, D. S., et al. 2003b. "Policies Promoting Innovation in Small Firms: Evidence from the U.S. and U.K." *Small Business Economics* 20 (2): 121–27.

Stankiewicz, R., 1986. *Academics and Entrepreneurs: Developing University-Industry Relations*. Pinter.

Thursby, J., A. W. Fuller, and M. Thursby. 2009. "US faculty patenting: Inside and outside the university." *Research Policy* 38 (1) 14–25.

Thursby, J. G., and S. Kemp. 2002. "Growth and productive efficiency of university intellectual property licensing." *Research Policy* 31 (1): 109–24.

Tijssen, R. J., and E. van Wijk. 1999. "In search of the European Paradox: An international comparison of Europe's scientific performance and knowledge flows in information and communication technologies research." *Research Policy* 28 (5): 519–43.

Valentin, F., and R. L. Jensen. 2007. "Effects on academia-industry collaboration of extending university property rights." *Journal of Technology Transfer* 32 (3): 251–76.

Verspagen, B. 2006. "University research, intellectual property rights and European innovation systems." *Journal of Economic Surveys* 20 (4): 607–32.

CHAPTER EIGHT

The Transition toward Entrepreneurial Universities

An Assessment of Academic Entrepreneurship in Italy

Nicola Baldini, Riccardo Fini, and Rosa Grimaldi

1. Introduction

Many universities are actively promoting academic entrepreneurship. Following Grimaldi et al. 2011, we define academic entrepreneurship as a university's engagement in the commercialization of research, including formal mechanisms such as academic start-ups, university patents, industry-university collaborations, and licensing (Baldini, Grimaldi, and Sobrero 2006; Fini, Lacetera, and Shane 2010; Phan and Siegel 2006; Lockett and Wright 2005; Siegel, Veugelers, and Wright 2007), as well as informal ones, such consulting and other networking, personnel-related activities with industrial partners (Perkmann et al. 2011; D'Este and Patel 2007; Perkmann and Walsh 2008).

Although several studies (e.g., Grimaldi et al. 2011; Wright et al. 2008) have improved our understanding of this phenomenon, it is far from clear how legislative and organizational changes interact with local context specificities and whether universities rooted in different cultures or economic contexts are likely to exhibit the same patterns of entrepreneurial activities. For a better understanding of academic entrepreneurship, we should adopt a comprehensive perspective, by including environmental trends, legislative frameworks, governmental policies, university level characteristics and policies, and by looking at how their inter relations and their evolution affect universities' ability to successfully engage in the commercialization of the research results. To this end, country-based case studies would be best suited, as long as they make it possible to gather fine-grained information and eventually to interpret them in the light of country specificities, including the legislative context, the organization of the public research system, and, more generally the historical setting.

Based on these premises, our chapter assesses academic entrepreneurship in Italy. As in many other countries, the transition toward commercialization of research results by Italian universities has been accompanied by legislative changes. In this study, we combine a characterization of the top-down amendments to the intellectual property rights (henceforth, IPR) regime in the light of the contemporaneous reforms occurring in other European countries with a comprehensive description of the way Italian universities responded in terms of internal policies and mechanisms in support of the commercialization of research results.

We also provide descriptive evidence on the amount of academic entrepreneurship (assessed in terms of number of patents, number of spin-off established and amount of consultancy), delivered by the 64 Italian Science, Technology, Engineering and Mathematics (STEM) universities since early 2000s.

The remainder of this chapter is organized as follows. In section 2, we describe the rationale for universities' transition toward academic entrepreneurship. Section 3 provides an overview of the European situation, in terms of IPR changes and university patenting. Section 4 introduces the Italian setting in terms of legislative context and universities' characterization. Section 5 presents empirical evidence on academic entrepreneurship during the previous decade at the 64 STEM universities in Italy. Section 6 provides conclusions.

2. Rationale behind Universities' Transition toward Academic Entrepreneurship

There are several reasons for university engagement in academic entrepreneurship. First, it relates to the additional role that universities could and should have (on top of research and teaching) in terms of contributing to the technological and economic development in their respective countries. The natural flow of university knowledge to the market place, in the form of new academic start-ups, university patents and licenses, and university-industry collaborations, can be inhibited by market failures. Market failure can arise because of asymmetry of information. Third parties—including financial investors, banks, and companies to set up collaborations with—might lack the expertise to assess the positive outcomes deriving from investing in a start-up or set up collaborations with academics and therefore overestimate the risk of the investment/effort (Black and Gilson 1998). In addition, academics willing to engage in technology transfer might lack the ability to communicate to third parties the value embedded in their technological knowledge or might not be willing to share too many details on their technologies simply because they fear leakage/dissemination of information they consider critical for the successful commercialization of their knowledge (Nerkar and Shane 2003). Adverse selection arises because it becomes difficult for external third parties to distinguish between technologies and valuable opportunities worth pursuing.

As a result, the quest for institutional intervention to mitigate such inefficiency arises. Universities can play a significant role in mitigating these inefficiencies (e.g., via the creation of technology transfer offices [TTOs]), by developing optimal patent, start-up and consultancy policies, and other internal regulations/mechanisms signaling more formalized processes. While these internal policies and mechanisms do not represent a sufficient condition for the successful commercialization of university research results, they are important signaling mechanisms for external parties and stakeholders.

In addition to overcoming possible market inefficiencies for the benefit of society, universities have more selfish motives for supporting academic entrepreneurship. Given that these activities can potentially generate additional revenue for the university and its faculty (Etzkowitz et al. 2000). This attitude reflects a generalized tendency for decreasing federal/public sources of founding for research, which characterized all major industrialized countries during the 1990s (Geuna 1998; Calderini, Garrone, and Sobrero 2003). It follows that technology transfer activities are systematically considered as necessary to defending universities' competitive position by creating new mechanisms for funding research activities. The involvement in academic entrepreneurship has different forms of pay-offs for universities. On the pure financial side, new sources of revenues are attached to the royalties on licensed technologies, to the possible sale of shares in academic spin-offs during their different rounds of financing or IPOs, to research contracts with companies. Such revenues can then be pooled at the central administration level to support additional research activities, to hire research assistants, organize (and participate in) events, build and remodel laboratories, acquire new infrastructure; or revenues can be channeled directly to specific units such as departments or research centers.

Another form of pay-off is the contribution of these activities to enhancing the attractiveness of studying at the university, thus increasing the opportunities to attract smart people interested in the commercialization of their research results (Florida 1999). This holds for talented researchers as well, who, in addition to the traditional outcomes of research activities (publications and/or participation to conferences), are given the possibility of increasing revenue, to capture the rents for their rare and valuable intellectual property (Zucker, Darby, and Brewer 1998). Outstanding scholars may give themselves the chance to exploit their abilities in new and different ways, to prove them in new—though related—activities thus satisfying their own personal motivations (Rod 2006). Finally, academic entrepreneurship might also contribute indirectly to strengthen a more general set of opportunities of interaction with the industrial world (Guy and Quintas 1995; Friedman and Silberman 2003). Employment opportunities are enhanced, thus positively affecting placement services. A greater involvement in the projects directly attached to the development and engineering of new products coming from research activities favors the diffusion of a greater experience of industrial applications and direct links with companies. Both can generate positive spillovers for universities in terms of teaching, access to equipment, industry know-how and technical advice (Owen-Smith and Powell 2001), and more generally on the institutional reputation (Moutinho, Fontes, and Godinho 2007).

3. Intellectual Property Regimes in Europe

3.1 IPR Systems in European Countries

IPR laws are quite varied in Europe, and there is concern that different national legislation regarding the ownership and exploitation of IPRs may create barriers to international collaborative research (Schmiemann and Durvy 2003). In many European countries, until the beginning of the twenty-first century, universities were exempt from standard IPR laws granting IPRs on employees' inventions to the employers. This is known as the so-called professor's privilege or teacher exemption clause (OECD 2003; Geuna and Rossi 2011). This label stems from the rule being applied only to researchers working in universities, but not to those working in public research organizations, which on the contrary retain entitlement to patents. A number of European countries changed employment or IPR laws so that university professors were no longer exempt from legislation giving employers IPRs generated by employees, and transferred ownership from individual inventors to universities.

According to Calderini, Garrone, and Sobrero (2003), Geuna (2001), and Mowery and Sampat (2005), rationales for these reforms included, among other things, a catalytic effect of the Bayh-Dole Act on university-industry technology transfer in the U.S.; a better use of research results generated with public funds than could otherwise remain unexploited; the creation of employment through academic spin-offs or start-ups, an increasing fund for public research in a period of shrinking budgets; the elimination of barriers to international collaborative research created by different national laws regarding the ownership and exploitation of IPRs. An overview of IPR ownership in major European countries is reported in table 8.1.

Despite the abolition of the professor's privilege, we are far from having a common European regime with respect to IPRs on employee's inventions: there are at least two broad families. When the inventive mission is expressly stated in the contract, the so-called service invention belongs to the employer originally through implied cession (available, e.g., in France, Greece, Netherlands, Portugal, and U.K.), and the inventor may or may not be entitled to compensation (e.g., if the invention is particularly profitable to the employer or if personal contribution exceeds the content of the employment contract). Preemption rights arise when the invention has some connection with the employer's activity (e.g., falls within the scope of the company's field of activity, or it is the result of an inventive mission entrusted to the employee): in such cases, the employer has the possibility (but not the obligation) to claim rights over the patent during a certain period of time, after which the IPRs revert to the inventor, which therefore is said to have a "right of first refusal" (available, e.g., in Denmark, Germany, Finland, and Norway); the employee cannot refuse claim, but deserves additional compensation. Additional provisions further complicate the issue. Last but not least,

TABLE 8.1. **University patenting activity: IPR-ownership.**

Country	University	Inventor
Austria	from October 1, 2002	(*)
Belgium (Flanders)	from January 1, 1998	
Denmark	from January 1, 2000	
Finland	from January 1, 2007	
France	1982	
Germany	from February 7, 2002	
Greece	1995	
Italy	before October 25, 2001, and after March 4, 2005	between October 25, 2001, and March 4, 2005
Netherlands	1995	
Norway	from January 1, 2003	
Spain	1986	
Sweden		1949
U.K.	1977	

(*) Until September 30, 2002, the Ministry of Science is the employer of university professors, and as such is entitled to IPRs on their inventions.

in some countries like Austria, Ireland, Italy, Sweden, and U.K., written agreements may even derogate national legislation on IPRs, and the position with respect to the ownership of IPR ultimately depends on the circumstances of each individual case.

3.2 Differences in University Patenting between Europe and U.S.

There have been numerous papers on university patenting in Europe. Some of these studies have pointed to differences between university patenting in Europe and in the U.S. (e.g., Jacob, Lundqvist, and Hellsmark 2003; Geoghegan and Pontikakis 2008; Wright et al. 2008).

First, cooperation between firms and universities is still insufficiently developed in the majority of European countries (de Juan 2002; Viale and Campodall'Orto 2002). Second, the European and U.S. patent systems are different. For example, the European patent system is not highly active, because of higher costs, the lack of a grace period (Geuna and Nesta 2006), and difficulties in raising funds for early-stage technologies and bringing them to market (Henrekson and Rosenberg 2001). Also, the non-patentability of software-related inventions is distorting competition among the E.U., Japan, and the U.S. (de Juan 2002).

Studies by Iversen, Gulbrandsen, and Klitkou (2007) on Norway, by Meyer (2003) on Finland, by Saragossi and van Pottelsberghe (2003) on Belgium, by Schmiemann and Durvy (2003) and Czarnitzki, Glänzel, and Hussinger (2007) on Germany, by Wallmark (1997) and Lissoni et al. (2008) on Sweden, indicate that many inventions are developed at universities, yet applied for by other institutions owing to different IPR rules (the so-called professor's privilege). Moreover, Balconi, Breschi, and Lissoni (2003) on Italy, Lissoni et al. (2008) on France and Italy, and Crespi, Geuna, and Nesta (2007) on six European countries stress that it is very common (up to 85% of the analyzed cases) not to transfer IPRs to the university, in violation of existing rules and laws, indicating that European universities might have a lower bargaining power with respect to industry, when compared to U.S. ones. At the same time, Thursby, Fuller, and Thursby (2009) find that a quarter of patents on which U.S. faculty are listed as inventors are assigned solely to firms rather than to the universities, as dictated by university policies or the Bayh-Dole Act. Valentin and Jensen (2007) offer an appealing quasi-controlled experiment by systematically comparing patent-related university-industry collaboration in Denmark and Sweden before and after the elimination of the professor's privilege in the first but not in the latter country in year 2000. The new IPR law in Denmark reduced contributions to Danish biotech firms' patents from Danish academic inventors, combined with a simultaneous substitutive increase of non-Danish academic inventors. The authors observe a moderate increase in academic inventions channeled into university-owned patents, but underscore that the larger part of the inventive potential of academia seems to have been rendered inactive as a result of the reform, probably owing to the high uncertainties and the exploratory nature of the research field under investigation (i.e., biotechnology).

Only a few studies compare patenting performance across countries (e.g., Meyer et al. 2005; Hehrer 2006; Rasmussen, Moel, and Gulbrandsen 2006; Decter, Bennett, and Leseure 2007; Wright et al. 2008), and fewer still examine contrasting performances using econometric models (e.g., Crespi, Geuna, and Nesta 2007; Valentin and Jensen 2007). More generally, econometric models outside the U.S. are rare, but their number is increasing (e.g., Breschi, Lissoni, and Montobbio 2008; Chapple et al. 2005; Azagra-Caro et al. 2006 and 2007; Calderini, Franzoni, and Vezzulli 2007). This paucity may also be the result of complete and regular surveys on patent and revenue data being scarce outside North America. In contrast, the survey by the U.S. and Canada professional association of technology transfer officials in academia, namely the Association of University Technology Managers (AUTM), is a widely used tool (e.g., Thursby and Thursby 2002; Friedman and Silberman 2003; Siegel, Waldman, and Link 2003; Sine, Shane, and Di Gregorio 2003; Chukumba and Jensen 2005). This survey is now replicated in other selected countries, such as in Denmark by Inside Consulting and subsequently the National Network for Technology Transfer from 2000 (e.g., Baldini 2006), in the U.K. by the University Companies Association from 2001 (e.g., Chapple et al. 2005; Hehrer 2006), and in Italy by the Network for the Valorization of University Research Results from 2002 (e.g., Baldini 2010). However, no cross-country surveys exist. The OECD's patenting and licensing survey (2003) was a commendable but one-time effort. It is also important to note that that these are sample and response biases that might seriously limit effective cross-country comparisons. Regular surveys of patenting and licensing activities are needed to provide input to policy makers but also to help universities benchmark performance and learn from one another (OECD 2003).

The OECD (2003) warned that, despite favorable rules on IPR ownership, university patenting activity may remain low because of insufficient incentives, beyond legal requirements, to exploit IPR, as well as barriers unrelated to IPR legislation. In the next section, we connect the university and other reforms with those on IPR legislation in the context of Italy.

4. Academic Entrepreneurship in Italy

The Italian university system has 94 public universities, 64 of them conducting research into Science, Technology, Engineering and Mathematics fields (STEM universities). Located in 19 of the 20 Italian regions[1], STEM universities employed, in 2007, an average of 890 academics. Until the late 1980s, as a result of the fully public and highly centralized characteristics of the Italian system, these universities had low levels of autonomy, having all the strategic decisions undertaken by the central government, in Rome.

In particular, the national government has significant power, in determining the allocation of financial resources, not only among universities (as is still the case), but also determining, within each institution, the allocation of resources among subject areas (e.g., how much should be spent on stationery and how much on routine building maintenance). Also, the recruitment and the promotion of teaching and nonteaching staff is determined by the central government.

The original IPR Code, dating back to 1939, made no distinction between workers in universities or public research organizations and other workers. If an invention is made in the course of an employment or work contract, the invention belongs to the employer[2]. If there is no specific remuneration for inventive activity (which is the case for universities/ public research organizations' employees), if the employer is granted the patent, then the inventor is entitled to equitable remuneration depending on the importance of the invention, on the duties and the income of the inventor, and on the help from the employer's company. Despite IPR ownership, universities lacked the financial autonomy to invest in the management of IPRs.

4.1 The University Reforms

Universities all report to the Ministry of Education, University and Research (MIUR), that was established (with the name of the Ministry of University and Scientific and Technological Research) as a result of an autonomous acquisition process beginning at the end of the 1980s with the enforcement of Laws n. 400/1988 (i.e., self-regulation principle) and n. 168/1989 (i.e., establishment for the first time of a ministry for university and research separated from that for education). From the early 1990s, the new ministry granted greater autonomy to academic institutions, allowing them to manage their budgets, to design their teaching programs, and to introduce statutes and regulations for managing organizational and scientific activities, locally. Universities therefore created internal mechanisms to support academic entrepreneurship, namely the exploitation of university-owned IPR, the creation of academic spinoffs and the consultancy activities of academics. The majority of Italian STEM universities put at place policies meant to efficiently regulate these activities.

The most important change for technology transfer was a national law creating the conditions necessary for the effective commercialization of research results through academic start-ups. D. Lgs.[3] n. 297/1999 specifically fostered the creation of research spin-offs, by introducing the possibility for public researchers to be formally involved in the creation of a spin-off or in other technology transfer projects between a university/PRO and a firm, while keeping their position and wage (up to eight years). However, this is not a leave of absence, as academics are not relieved of their teaching duties. In order to foster an entrepreneurial shift, this law also provides financial benefits to research spin-offs. Following the introduction of university autonomy, the practical implementation of the new law has been left to each institution, to elaborate its own regulations and units to manage the spin-off matter.

At the level of each university, patent, spin-off, and consultancy policies are intended to rule the commercialization activities of academics, having two main goals. First of all, they clearly define the rights of all parties involved in the transaction on both the academic and the industrial side, as well as their remunerations (if applicable). Specifically, the patenting policy states to whom the invention must be disclosed and who is entitled to patent, whereas the spin-off regulation describes how and to what extent academics can get involved in corporate roles, and, finally, the consultancy policies characterize the appropriability regime of the IPR stemming from the collaboration. Secondly, all of these policies govern a university's involvement in and support for the technology-transfer activity, resulting in legal, financial, and marketing support for the individuals involved in the process.

Until the beginning of the last decade, besides the aforementioned efforts, there were no formal associations to support and advance academic technology transfer in the national context, such as AUTM in the U.S. (www.autm.net) and AURIL in the U.K. (www.auril.org.uk). Only in 2002, as a result of an initiative of the Polytechnic of Milan, was the Network for the Valorisation of University Research (NetVal) created (www.netval.it). NetVal was established to support and diffuse technology commercialization activities in academic institutions. An agreement protocol was signed in 2001 by 29 Italian universities, which grew to 46 in 2009. Since inception, the network has offered 34 courses, and trained about 1,000 individuals (mostly administrative and technical university employees) on IPR-related matters.

4.2 The IPR Reforms

In 2001, the Law n. 383 amended the IPR code and introduced the professor's privilege, thus transferring the rights in the opposite direction in respect to the legislative reforms that occurred in other European countries since the millennium. Another difference was that the professor's privilege applied to all workers in universities, and to those in public organizations having research as a part of their mission. The research institutions, however, secured a large monetary compensation, and were free to choose its amount, not exceeding half of the net revenues stemming from licensing activity (if the choice of amount is not made, the institution is automatically entitled to 30% of the net revenues); moreover, the research institution was granted compulsory nonexclusive licenses to prevent failure to work or insufficient working for five years after the date of the patent.

Since the introduction of the 2001 Law, a heated debate has ensued, involving different actors, including industry associations, universities and public research organizations, as well as political parties. Despite their different mandates, they unanimously called for the elimination of such a law, claiming that it discriminated between private and public employees, increased complexity and uncertainty in IPR negotiations on jointly private-public projects, and provided no incentives to universities and public research organizations to strategically manage inventions developed in their labs. As a result of such institutional pressures, the national government recognized that public researchers had rarely patented and economically exploited their inventions, and issued a new IPR Code (D.Lgs. n. 10/2005 of February 10, 2005). Following the new legislation, the provision introduced in 2001 did not apply when inventions stemmed from research that was at least partially privately financed or from specific research projects that were financed by public institutions different from the one to which the inventors belonged. It is important to underscore that, in Italy, such cases are the rule rather than the exception.

Overall, the IPR legislation in Italy has been extremely uncertain over the last 15 years. For example, a proposal to abolish the 2001 law was approved by the Senate as early as January 29, 2002, but was still pending at the Chamber of Deputy in September 2002. More recently, the draft for the first amendment to the new IPR code (D. Lgs. n. 131/2010) originally meant to totally abolish the professor's privilege (where still at place) and converting it to a right of first refusal for the inventors. This new rule, however, disappeared from the version of the law that was finally approved on August 13, 2010.

5. Empirical Evidence on Academic Entrepreneurship at Italian Universities

In order to assess the level of academic entrepreneurship by the Italian STEM universities, we examined the (annual) number of patent families[4] granted to academic institutions, the number of academic spin-offs established, and the monetary value of consultancy activities undertaken by academics. Patents information was collected using both the PATUNIT (Baldini, Grimaldi, and Sobrero 2006) and ORBIT (www.orbit.it) databases. For the number of spin-off companies, we relied on information stored in the IRIS-database[5] (Fini et al. 2011), while the amount of consultancies were downloaded from the MIUR website (www.miur.it). Figure 8.1 reports the trends for the selected variables, showing over time, a substantial increase of both patenting and consultancy activities, as well as a constant pattern in the number of spin-offs established.

In order to estimate an aggregate indicator of academic entrepreneurship by Italian STEM universities, we performed a principal component analysis of the three aforementioned dimensions. Specifically, we first pooled all the observations over the period covered by our analysis. Secondly, given the different nature of the selected variables, we standardized all three indicators included in our exercise. As the loading threshold for component identification we referred to 0.5 (Hair et al. 1995) and adopted a Kaiser-Meyer-Olkin measure of sampling adequacy level of 0.6 (Tabachnik and Fidell 2001). As reported in table 8.2, all items loaded on a single principal component with factor loadings higher than 0.50. The proportion of variance explained was almost 55%, with a sampling adequacy level of 0.61. Table 8.3 exhibits the descriptive statistics for the three items as well as for the principal component.

Figure 8.2 reports the percentage of STEM universities that adopted a patent, a spin-off or a consultancy policy, in the 2001–2007 period. In 2000, none had a spin-off policy, while only a few had a patent (about

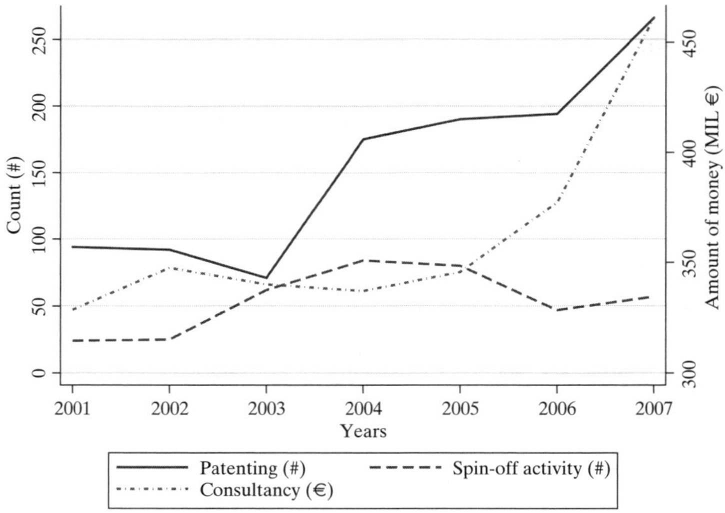

FIGURE 8.1. Academic entrepreneurship in the Italian STEM university.
Source: for patenting, PATUNIT (Baldini et al., 2006) and ORBIT (www.orbit.it); for spin-off activity, IRIS-Database (Fini et al., 2011); for consultancy, MIUR (ww.miur.it).
Notes: Where more than one university is among the assignees, we assigned the patent to each of them. Monetary values are all reported in 2001 currency.

TABLE 8.2. **Principal component analysis.**

Variable	Academic entrepreneurship (principal component)
Spin-off activity	0.61
Patenting	0.60
Consultancy	0.52
Proportion of variance explained (%)	54.55
Kaiser-Meyer-Olkin measure of sampling adequacy level	0.61

TABLE 8.3. **Academic entrepreneurship in Italy (2001–2007).**

Variable	Observations	Mean	Std. Dev.	Min	Max
Spin-off activity (#)	448	0.85	1.60	0	10
Patenting (#)	448	2.42	4.19	0	26
Consultancy (€)	370	6,857,199	8,526,326	846	39,100,000
Academic entrepreneurship (principal component)	370	0.00	1.27	-3.17	5.72

Note: Information on consultancy activities is missing in 78 cases.

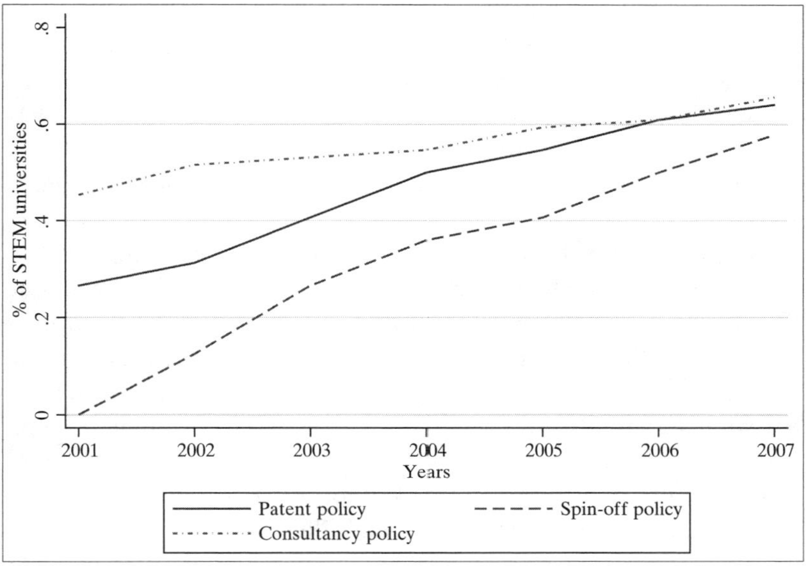

FIGURE 8.2. Percentage of STEM universities with patent, spin-off, and consultancy policies.

30%) or a consultancy (about 45%) regulation. On the contrary, in 2007, almost 60% of them had the patent, the spin-off or the consultancy polices in place (however, only 30% of them had all three policies in place at the same time; please refer to figure 8.3 for more detailed information).

We then assessed trends in academic entrepreneurship by the Italian STEM universities, in the 2001–2007 period. Moreover, we looked for differential trends in academic entrepreneurship as a result of the presence of specific support mechanisms. Specifically, in figure 8.3 we compared whether or not universities had all three policies in place, an affiliation to NetVal, an operational TTO, and invested in the professionalization of the TTO personnel[6].

As the solid line reported in figure 8.3a shows, the level of academic entrepreneurship increased significantly between 2001 and 2004 then flattened in the following years. Comparing universities having all three policies in place with those that do not, our results suggest that the former deliver markedly higher levels of academic entrepreneurship compared to the latter. However, data also shows that the amount of academic entrepreneurship by universities with all three policies in place decreases over time after 2005; this is a result of the lower performance of the ten universities that adopted all three polices in later years, between 2004 and 2007 (as showed by the dashed line in the figure). Data also shows that, in 2007, less than 33% of Italian STEM universities adopted all three policies at the same time. The average trends in Italian academic entrepreneurship are therefore characterized—to a large extent—by universities without the three policies (as showed by dash-dotted line in the figure).

We proceeded with our analysis by comparing the performance of universities affiliated to NetVal with those without such affiliation. Further differences also emerged, with affiliated universities outperforming the rest over the entire period (see figure 8.3b). Despite a drop in the number of affiliations in 2006, academic entrepreneurship by universities with NetVal affiliation has constantly increased overtime, peaking in 2007.

In a similar fashion, we also looked for differences between universities with and without TTOs. The trends illustrated in figure 8.3c show that, over time, an increasing number of universities have established a TTO. Only 35 universities had a TTO in 2004 (slightly more than 50% of the population), whereas, in 2007 this figure had risen to 55 (more than 85% of the population). As the average level of academic entrepreneurship remained constant over the whole period, the trend shows a significant drop in the 2007 performance of the few STEM universities without a TTO.

Finally, we compared universities with at least one TTO employee attending Net Val courses. The evidence reported in figure 8.3d shows that about 40% of STEM universities invested in the professionalization of their TTO employee, while the remaining 60% did not. As already recorded above, the former group outperformed the latter. Figure 8.3 reports this trend.

We then assessed the value of academic entrepreneurship at each STEM institution, aggregating them by region. Italy is divided into 20 regions, all ruled by the central government. Regions are the first-level administrative divisions of the Italian state. As table 8.4 reports, STEM universities are uniformly distributed over the country. There are, on average, about three STEM universities per region, with Lombardy, the region in which Milan is located, being the densest with ten universities. As for academic entrepreneurship, Emilia Romagna is the best performer in terms of both total (33.9) and average (1.26) values, followed by Piedmont (17.34) and Tuscany (16.76), for total values, and by Piedmont

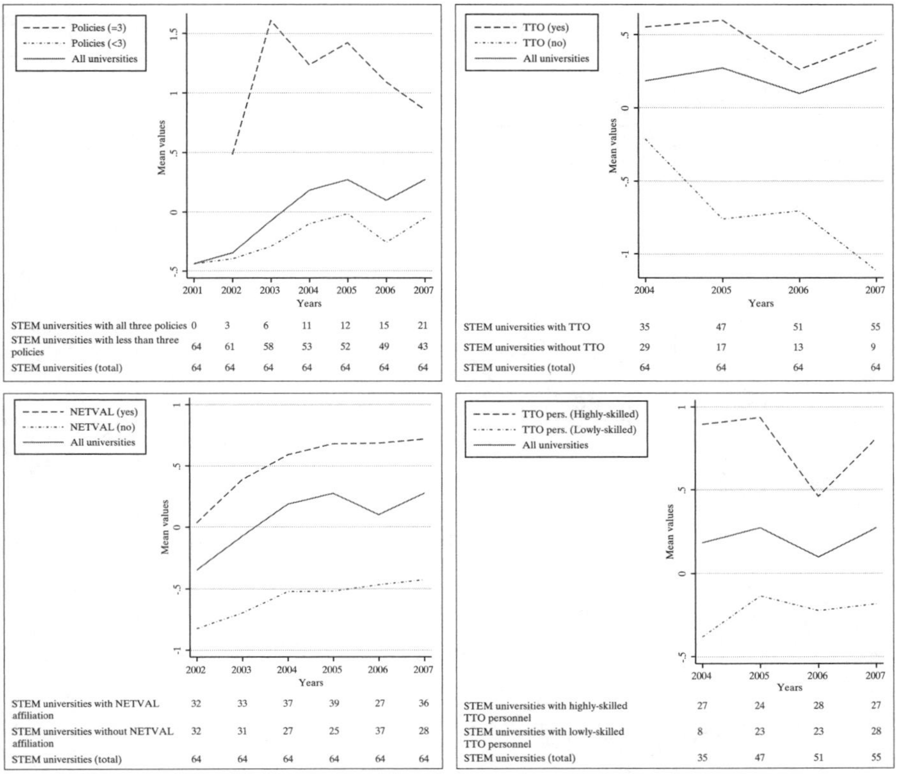

	2001	2002	2003	2004	2005	2006	2007
STEM universities with all three policies	0	3	6	11	12	15	21
STEM universities with less than three policies	64	61	58	53	52	49	43
STEM universities (total)	64	64	64	64	64	64	64

	2004	2005	2006	2007
STEM universities with TTO	35	47	51	55
STEM universities without TTO	29	17	13	9
STEM universities (total)	64	64	64	64

	2002	2003	2004	2005	2006	2007
STEM universities with NETVAL affiliation	32	33	37	39	27	36
STEM universities without NETVAL affiliation	32	31	27	25	37	28
STEM universities (total)	64	64	64	64	64	64

	2004	2005	2006	2007
STEM universities with highly-skilled TTO personnel	27	24	28	27
STEM universities with lowly-skilled TTO personnel	8	23	23	28
STEM universities (total)	35	47	51	55

FIGURES 8.3A–D. Trends in Italian academic entrepreneurship.
Note: The number of universities affiliated to NETVAL decreases in 2006 because of an organizational restructuring. NETVAL, indeed, stopped being an informal network, becoming a formal association.
Source: IRIS-Database (Fini et al., 2011)

(0.91) and Veneto (0.51), for mean figures. Campania (5) and Molise (12) are the laggards in terms of total (–20.43) and average (–1.26) amount of academic entrepreneurship.

Figures 8.4 and 8.5 report a visual representation of academic entrepreneurship (mean and total values) delivered by universities located in each region. Universities located in northern and central regions consistently outperform the southern ones in terms of both total and mean values. No differences in regional performance are reported for both total and average values, with the only exceptions being universities located in Lombardy (10) and Basilicata (3), that show better performance in terms

TABLE 8.4. **Academic entrepreneurship by region.**

Region (name)	Region (id)	STEM universities	Number of observations (2001–2007)	Number of missing values	Academic entrepreneurship (principal component) (mean values)	Academic entrepreneurship (principal component) (total values)
Abruzzo	1	3	21	5	−1.14	−18.19
Apulia	2	4	28	1	−0.23	−6.33
Basilicata	3	1	7	0	−0.66	−4.63
Calabria	4	3	21	0	−0.62	−12.99
Campania	5	5	35	2	−0.62	−20.43
Emilia-Romagna	6	4	28	1	1.26	33.90
Friuli-Venezia Giulia	7	3	21	5	−0.45	−7.24
Lazio	8	6	42	12	−0.22	−6.70
Liguria	9	1	7	0	0.11	0.74
Lombardy	10	10	70	27	0.38	16.40
Marche	11	3	21	6	−0.22	−3.34
Molise	12	1	7	0	−1.26	−8.84
Piedmont	13	3	21	2	0.91	17.34
Sardegna	14	2	14	2	−0.01	−0.15
Sicily	15	4	28	8	−0.58	−11.58
Tuscany	16	5	35	0	0.48	16.76
Trentino–Alto Adige	17	2	14	7	0.29	2.02
Umbria	18	1	7	0	0.38	2.66
Veneto	20	3	21	0	0.51	10.62
Total		64	448	78		

Note: Valle d'Aosta (19) has no STEM universities.

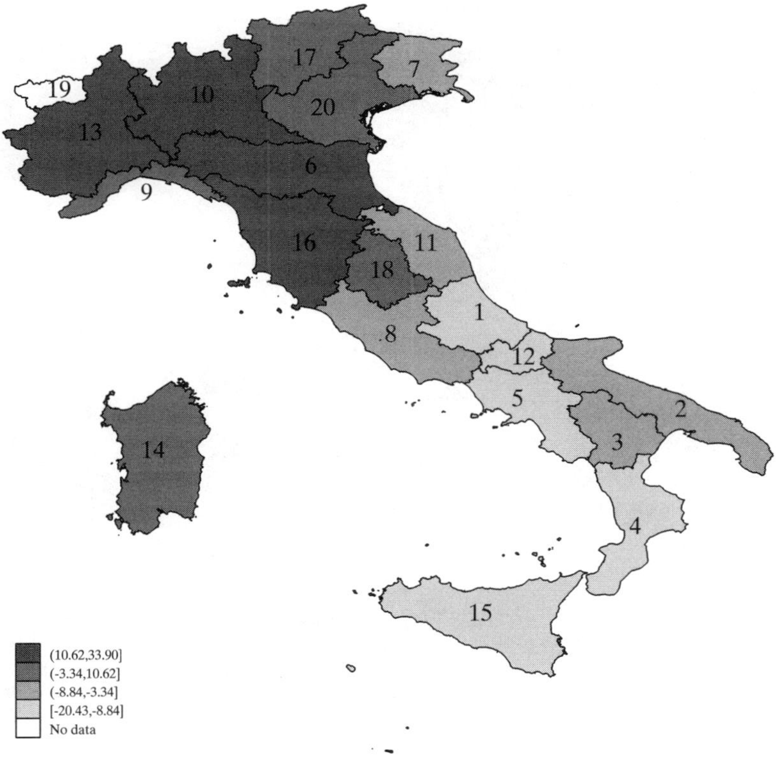

FIGURE 8.4. Total value of academic entrepreneurship by region.
Source: IRIS-Database (Fini et al. 2011)

of total rather than mean values, and institutions located in Veneto (20) and Sicilia (15), recording better performance in terms of mean rather than absolute figures.

Finally, in order to provide more information on the changes in academic entrepreneurship over the 2001–2007 period, figure 8.6 plots the mean level of academic entrepreneurship in 2001 against the change of this variable over the following six years (2001–2007). Each dot represents the mean value of academic entrepreneurship generated by the STEM universities localized in any given region. The two dashed lines exhibit the average values for the plotted variables.

Figure 8.6 shows that academic entrepreneurship increased in all regions, with changing rates comprised between 0.8 (Trentino Alto Adige) and 2.35 (Apulia). The top left quadrant includes the initially advanced

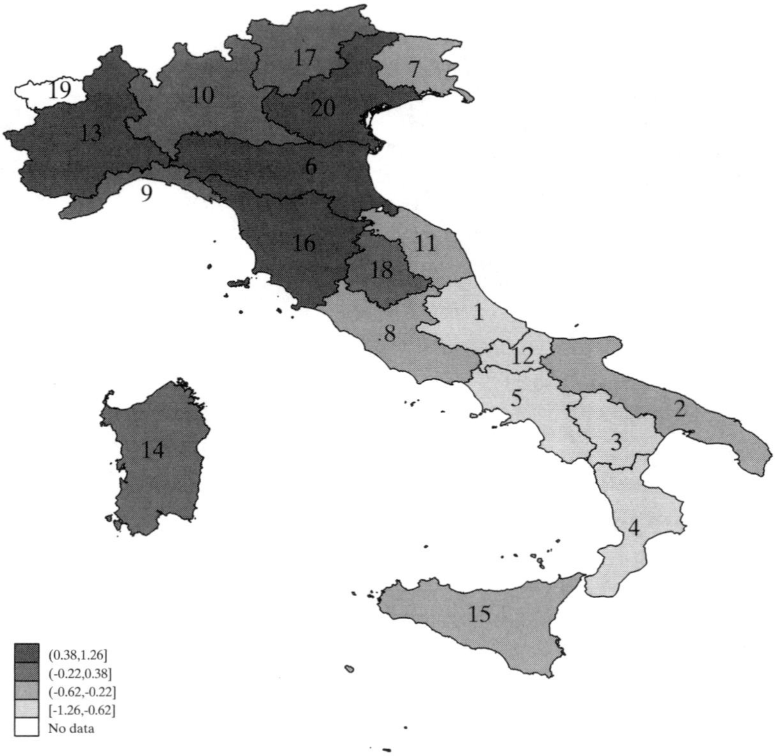

FIGURE 8.5. Mean values of academic entrepreneurship by region.
Source: IRIS-Database (Fini et al. 2011)

regions that grew slowly in the 2001–2007 time period. This group includes, for example, universities localized in Emilia Romagna and Toscana, which have been among the first to implement the normative and infrastructural mechanisms to support academic entrepreneurship. More specifically, in the late 1990s, the University of Bologna and Sant'Anna Pisa, together with the Polytechnic of Milan located in Lombardy have been among the pioneers in formally regulating university-patenting activities (Baldini et al., forthcoming).

The bottom right section refers instead to the initially lagging regions that are catching up. In this quadrant we find the universities located in Apulia and Sardegna, which have recently put a lot of effort in creating a favorable context for entrepreneurship. The ten STEM universities located in Lombardy were also included in this section. Indeed, their performance has been quite heterogeneous, with some institutions, such as the Polytechnic of Milan, outperforming the vast majority of Italian institutions, but also with some others falling behind.

Although entrepreneurial efforts are increasing—on average—for all universities throughout the 20 regions, the bottom left quadrant shows a number of lagging regions, all localized in the south and center of Italy, delivering performance below average.

Finally, in the top right quadrant there are only two (virtually one) regions, for which the universities—that were already advanced in 2001—are continuing to pull ahead. Indeed, the universities located in Piedmont, led by the performance of the Polytechnic of Turin, recorded higher level of academic entrepreneurship than Italian averages in terms of both 2001 value (0.26) and 2001–2007 changes (1.88). Indeed, Piedmont universities were among the top three in 2001 and are among the top five in terms of growth rates.

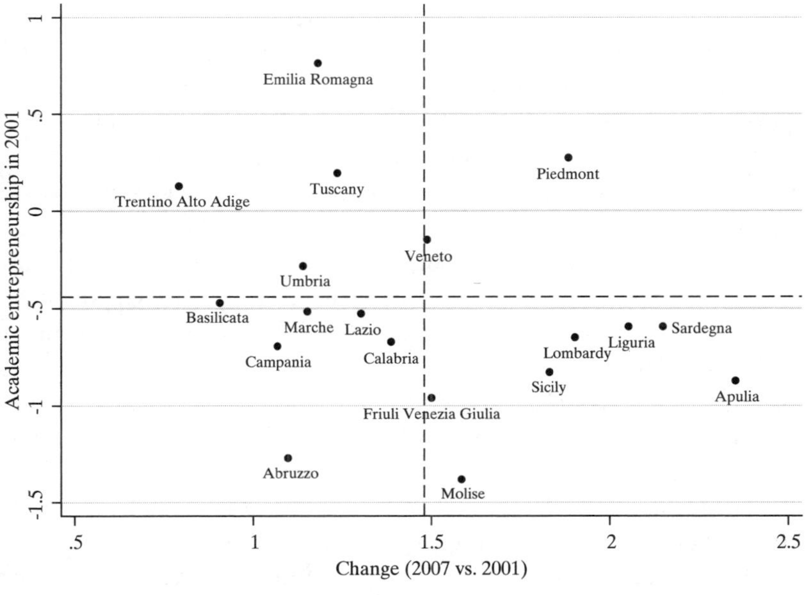

FIGURE 8.6. Changes in mean values of academic entrepreneurship by region. Source: IRIS-Database (Fini et al. 2011)

6. Conclusions

Our study shows the proactive role of Italian universities in stimulating academic entrepreneurship. We have examined the relationship between an aggregated indicator of academic entrepreneurship and a series of support mechanisms implemented by Italian STEM universities, including patent regulations, spin-off regulations, collaboration regulations, TTOs, and affiliation to NetVal—a university network to support and diffuse technology commercialization activities. The evidence shows that the three indicators of academic entrepreneurship activities are strongly positively correlated.

As for internal regulation, our data shows that universities with three internal regulations relating to patents, spin-offs, and collaborations have better results, in terms of academic entrepreneurship. This result holds true until 2005. After this year, while the number of universities adopting the three regulations increases, their average productivity in terms of academic entrepreneurship decreases. However, their productivity is significantly higher than that delivered by universities without all three policies in place. A potential explanation is that adoption of the three regulations merely reflects an isomorphic process, rather than the willingness to actively engage in entrepreneurial activity. This may have spurred universities to adopt internal mechanisms in order to be legitimized in a context where technology transfer activities from universities are socially desirable. In this regard, Baldini et al. (2010), while analyzing the diffusion and adoption of patent regulations by Italian universities, found that, at the time when Italian universities were granted greater autonomy in several areas, including the management of IPRs, many were uncertain as to what to do, and preferred to wait for the actions by three of the most prestigious institutions, replicating their behaviors and taking them as role models. This evidence supports the idea that academic institutions, in order to deal with normative changes, mimetically replicated the behaviors of the most legitimated organizations.

The availability of TTOs slightly affects the level of academic entrepreneurship. Though the presence of internal TTO personnel ad hoc trained for technology transfer positively influences academic entrepreneurship. The creation of internal TTOs, and the adoption of regulations, might reflect the intention by some universities to conform to a general trend according to which universities' involvement in the commercialization of research results is socially desirable. However, the availability of TTO personnel trained ad hoc reflects the willingness of TTOs to develop specific competences to successfully support technology transfer activities.

Universities in northern Italy perform better than universities in southern Italy, ceteris paribus. This suggests the importance of the external environment in fostering the commercialization of research results by universities. In a recent paper, Fini et al. (2011) analyzed academic spin-offs in Italy, reporting that university-level support mechanisms complement the legislative support offered to high-tech entrepreneurship at regional level. Their findings support the belief that the idiosyncrasies of regional settings should be taken into account in order for universities to develop effective spin-off-support policies.

Our analysis also shows that some "late mover" universities are trying to catch up with other universities more proactive in technology transfer. Most of the first mover universities having a rapid increase in academic entrepreneurship in the 1990s, have been slowing down in the period 2001–2006. One exception is represented by the Polytechnic of Torino. It is difficult to explain why this happens, without taking into account environmental factors, legislative changes, and also regional specificities. Our guess is that many first mover universities have faced more uncertainties than others, which, in most cases, has drained internal energies in the attempt to find the best internal configuration (these universities have also been adopting different version of patent and spin-off regulations over time).

Finally, the elaboration of an aggregate indicator of academic entrepreneurship was possible thanks to the IRIS data set. This file contains detailed data on university patents, spin-off companies, and university-industry collaborations. This is a very rich data set, which has been used for descriptive purposes in this chapter. In future, we intend to examine these data, as part of a more detailed analysis of academic entrepreneurship, in order to determine how individual, organizational, and institutional factors influence entrepreneurial success in an academic context.

Notes

1. Valle d'Aosta is the only Italian region without any STEM university.
2. The term "employer" is generically used in this chapter to indicate the part

who pays for the work, irrespective of the label attached to the part that performs the work (e.g., consultant, contractor, employee).

3. The legislative function in Italy is usually a prerogative of the Parliament. Sometimes, the Parliament can delegate the Government to exercise the legislative function in a given period of time, if the matter is thought to be particularly difficult and relevant. In such a case, the resulting norm is called *Decreto Legislativo* (abbreviated in D.Lgs.) instead of *Legge* (Law).

4. On the basis of a regular application for a patent filed by a given applicant in one of the 148 contracting states of the Patent Cooperation Treaty, the same applicant may, within twelve months, apply for protection in all the other member countries. These later applications will then be regarded as if they had been filed on the same day as the earliest application. All patents claiming the same priority are called a patent family, and count as one when patent counts are presented in this chapter.

5. The IRIS-database (Fini et al. 2011) was created in 2006 and, since then, it has been updated on yearly basis. It includes longitudinal information on the population of more than 400 academic spin-offs established in Italy during the last decade. It also comprises longitudinal information on the population of 64 STEM universities as well as of the 19 Italian regions in which they're localized in.

6. All variables were positively correlated, with the highest correlation coefficient being $r=0.48$; $p<.01$, assessed between the TTO presence and the professionalization of the its employees.

References

Azagra-Caro, J. M., et al. 2006. "Faculty support for the objectives of university-industry relations versus degree of R&D cooperation: The importance of regional absorptive capacity." *Research Policy* 35: 37–55.

Azagra-Caro, J. M., F. Archontakis, and A. Yegros-Yegros. 2007. "In which regions do universities patent and publish more?" *Scientometrics* 70: 251–66.

Balconi, M., S. Breschi, and F. Lissoni. 2003. "Il trasferimento di conoscenze tecnologiche dall'università all'industria in Italia: Nuova evidenza sui brevetti di paternità dei docenti." In *Il sistema della ricerca pubblica in Italia*, edited by A. Bonaccorsi, 58–100. Milan: Franco Angeli.

Baldini, N. 2006. "The act on inventions at public research institutions: Danish universities' patenting activity." *Scientometrics* 69: 387–407.

Baldini, N. 2010. "University spin-offs and their environment." *Technology Analysis and Strategic Management* 22: 859–76.

Baldini N., R. Grimaldi, and M. Sobrero. 2006. "Institutional changes and the commercialization of academic knowledge: A study of Italian universities' patenting activities between 1965 and 2002." *Research Policy* 35: 518–32.

Baldini, N., et al. 2014. "Organizational change and the institutionalisation of university patenting activity in Italy." *Minerva* 52 (1): 27–53.

Black, B., and R. Gilson. 1998. "Venture capital and the structure of capital markets: Banks versus stock markets." *Journal of Financial Economics* 47: 243–77.

Breschi, S., F. Lissoni, and F. Montobbio. 2008. "University patenting and scientific productivity: A quantitative study of Italian academic inventors." *European Management Review* 5: 91–109.

Calderini, M., C. Franzoni, and A. Vezzulli. 2007. "If star scientists do not patent: The effect of productivity, basicness and impact on the decision to patent in the academic world." *Research Policy* 36: 303–19.

Calderini, M., P. Garrone, and M. Sobrero. 2003. *Corporate Governance, Market Structure and Innovation*. Cheltenham, U.K.: Edward Elgar.

Chapple, W., et al. 2005. "Assessing the relative performance of U.K. university technology transfer offices: Parametric and non-parametric evidence." *Research Policy* 34: 369–84.

Chukumba, C., and R. Jensen. 2005. "University invention, entrepreneurship, and start-ups." Working paper no. 11475. Cambridge, MA: National Bureau of Economic Research.

Crespi, G. A., A. Geuna, and L. Nesta. 2007. "The mobility of university inventors in Europe." *Journal of Technology Transfer* 32: 195–215.

Czarnitzki, D., W. Glänzel, and K. Hussinger. 2007. "Patent and publication activities of German professors: An empirical assessment of their co-activity." *Research Evaluation* 16: 311–19.

Decter, M., D. Bennett, and M. Leseure. 2007. "University to business technology transfer: UK and USA comparisons." *Technovation* 27: 145–55.

de Juan, V. 2002. "Comparative study of technology transfer practices in Europe and the United States." *Journal of the Association of University Technology Managers* 14: 31–58.

D'Este P., and P. Patel. 2007. "University–industry linkages in the UK: What are the factors determining the variety of interactions with industry?" *Research Policy* 36 (9): 1295–1313.

Etzkowitz, H., et al. 2000. "The future of the university and the university of the future: Evolution of ivory tower to entrepreneurial paradigm." *Research Policy* 29: 313–30.

Fini, R., et al. 2011. "Complements or Substitutes? The Role Of Universities and Local Context in Supporting the Growth of Academic Spin-Offs." *Research Policy* 40: 1113–27.

Fini R., N. Lacetera, and S. Shane. 2010. "Inside or outside the IP-system: Business creation in Academia." *Research Policy* 39: 1060–69.

Florida, R. 1999. "The role of the university: Leveraging talent, not technology." *Issues in Science and Technology* 15 (4): 67–73.

Friedman, J., and J. Silberman. 2003. "University technology transfer: Do incentives, management, and location matter?" *Journal of Technology Transfer* 28: 17–30.

Geoghegan, W., and D. Pontikakis. 2008. "From ivory tower to factory floor? How universities are changing to meet the needs of industry." *Science and Public Policy* 35: 462–74.

Geuna, A. 1998. "The internationalization of European universities: A return to medieval roots." *Minerva* 36: 253–70.

Geuna, A. 2001. "The changing rationale for university research funding: Are there negative unintended consequences?" *Journal of Economic Issues* 35: 607–32.

Geuna, A., and L. Nesta. 2006. "University patenting and its effects on academic research: The emerging European evidence." *Research Policy* 35: 790–807.

Geuna, A., and F. Rossi. 2011. Changes to University IPR regulations in Europe and the impact on academic patenting. Research Policy 40(8), 1068–1076.

Grimaldi, R., et al. 2011. "Special Issue: 30 Years after Bayh-Dole: Reassessing Academic Entrepreneurship." *Research Policy* 40 (8): 1045–44.

Guy, K., and P. Quintas. 1995. "Collaborative, pre-competitive R&D and the firm." *Research Policy* 24: 325–48.

Hair, J., et al., eds. 1995. *Multivariate Data Analysis with Readings*. 4th ed. Upper Saddle River, NJ: Prentice-Hall.

Heher, A. D. 2006. "Return on investment in innovation: Implications for institutions and national agencies." *Journal of Technology Transfer* 31: 403–14.

Henrekson, M., and N. Rosenberg. 2001. "Designing efficient institutions for science-based entrepreneurship: Lesson from US and Sweden." *Journal of Technology Transfer* 26: 207–31.

Iversen, E. J., M. Gulbrandsen, and A. Klitkou. 2007. "A baseline for the impact of academic patenting legislation in Norway." *Scientometrics* 70: 393–414.

Jacob, M., M. Lundqvist, and H. Hellsmark. 2003. "Entrepreneurial transformations in the Swedish university system: The case of Chalmers University of Technology." *Research Policy* 32: 1555–68.

Lissoni F., et al. 2008. "Academic patenting in Europe: New evidence from the KEINS database." *Research Evaluation* 17: 87–102.

Lockett, A., and M. Wright. 2005. "Resources, capabilities, risk capital and the creation of university spin-out companies." *Research Policy* 34: 1043–57.

Meyer, M. S. 2003. "Academic patents as an indicator of useful research? A new approach to measure academic inventiveness." *Research Evaluation* 12: 17–27.

Meyer, M. S., et al. 2005. "Inventive output of academic research: A comparison of two science systems." *Scientometrics* 63: 145–61.

Moutinho, P. S. F., M. Fontes, and M. M. Godinho. 2007. "Do individual factors matter? A survey of scientists' patenting in Portuguese public research organizations." *Scientometrics* 70: 355–77.

Mowery, D. C., and B. N. Sampat. 2005. "The Bayh-Dole Act of 1980 and university-industry technology transfer: A model for other OECD governments?" *Journal of Technology Transfer* 30: 115–27.

Nerkar, A., and S. Shane. 2003. "When do start-ups that exploit patented 1010 academic knowledge survive?" *International Journal of Industrial Organization* 21: 1391–1411.

Organisation for Economic Co-operation and Development (OECD). 2003. *Turning Science into Business. Patenting and Licensing at Public Research Organizations*. Paris: OECD Publications.

Owen-Smith, J., and W. W. Powell. 2001 "To patent or not: Faculty decisions and institutional success at technology transfer." *Journal of Technology Transfer* 26: 99–114.

Perkmann, M., and K. Walsh. 2008. "Engaging the scholar: Three forms of academic consulting and their impact on universities and industry." *Research Policy* 37 (10): 1884–91.

Perkmann, M., et al. 2011. "Universities and the third mission: A systematic review of research on external engagement by academic researchers." Imperial College Working Paper.

Phan, P., and D. S. Siegel. 2006. "The effectiveness of university technology transfer: Lessons learned, managerial and policy implications, and the road forward." *Foundations and Trends in Entrepreneurship* 2 (2): 77–144.

Rasmussen, E., Ø. Moen, and M. Gulbrandsen. 2006. "Initiatives to promote commercialization of university knowledge." *Technovation* 26: 518–33.

Rod, M. 2006. "The sealing of university intellectual property boundaries and the ceiling of academic entrepreneurial tolerance." *Science and Public Policy* 33: 3–4.

Saragossi, S., and B. van Pottelsberghe de la Potterie. 2003. "What patent data reveal about universities: The case of Belgium." *Journal of Technology Transfer* 28: 47–51.

Schmiemann, M., and J.-N. Durvy. 2003. "New approaches to technology transfer from publicly funded research." *Journal of Technology Transfer* 28: 9–15.

Siegel, D. S., D. A. Waldman, and A. N. Link. 2003. "Assessing the impact of organizational practices on the productivity of university technology transfer offices: An exploratory study." *Research Policy* 32: 27–48.

Siegel, D., D. Veugelers, and M. Wright. 2007. "University commercialization of intellectual property: Policy implications." *Oxford Review of Economic Policy* 23 (4): 640–60.

Sine, W. D., S. Shane, and D. Di Gregorio. 2003. "The halo effect and technology licensing: The influence of institutional prestige on the licensing of university inventions." *Management Science* 49: 478–96.

Tabachnik, B. G., and L. S. Fidell, eds. 2001. *Using multivariate statistics (4th ed.)*. Needham Heights, MA: Allyn and Bacon.

Thursby, J. G., A. Fuller, and M. C. Thursby. 2009. "US faculty patenting: Inside and outside the university." *Research Policy* 38 (1): 14–25.

Thursby, J. G., and M. C. Thursby. 2002. "Who is selling to the ivory tower? Sources of growth in university licensing." *Management Science* 48: 90–104.

Valentin, F., and R. L. Jensen. 2007. "Effects on academia-industry collaboration of extending university property rights." *Journal of Technology Transfer* 32: 251–76.

Viale, R., and S. Campodall'Orto. 2002. "An evolutionary Triple Helix to strengthen academy-industry relations: Suggestions from European regions." *Science and Public Policy* 29: 154–68.

Wallmark, J. T. 1997. "Inventions and patents at universities: The case of Chalmers University of Technology." *Technovation* 17: 127–39.

Wright, M., et al. 2008. "Mid-range universities' linkages with industry: Knowledge types and the role of intermediaries." *Research Policy* 37: 1205–23.

Zucker, L. G., M. R. Darby, and M. B. Brewer. 1998. "Intellectual human capital and the birth of U.S. biotechnology enterprises." *American Economic Review* 88: 290–306.

CHAPTER NINE

Northeastern University

A Study of Technology Transfer and the Academic Entrepreneur

Tucker J. Marion, Denise Dunlap, and John H. Friar

* This paper constitutes an updated, extended version of an earlier manuscript by T. Marion, D. Dunlap, and J. Friar (2012), "The university entrepreneur: A census and survey of attributes and outcomes," *R&D Management* 42 (5): 401–19.

Introduction

Since the 1980s, research universities have been increasingly viewed by policy makers as engines of economic growth, via the commercialization of intellectual property through technology transfer into the private sector (Markman et al. 2005; Markman, Siegel, and Wright 2008; Siegel and Phan 2004)—going directly from primary research to commercialized innovation. In 1980, the Bayh-Dole Act was passed and since then there have been approximately 5,700 university-based start-ups formed.

The translation of technology into high-growth new, stand-alone businesses or business units is paramount to sustaining U.S. economic growth and standards of living. Research suggests, however, that only 12 to 16% of university-assigned inventions are transferred to private new ventures (Di Gregorio and Shane 2003; AUTM 2001 and 2002). The current literature on technological knowledge transfer efficacy has been primarily focused on the generation of patents and licensing (Siegel and Phan 2004), and not on the key component needed to foster development of sustainable new ventures, the academic entrepreneur. This research focuses on university entrepreneurship and on those individuals who perform the science there and are central figures in seeing that technology commercialized.

Mosey and Wright (2007) note that there have been few studies of academic entrepreneurs. According to Markman, Siegel, and Wright (2008), the process by which academic inventions are transferred or spill over into external benefits such as new start-ups outside of the university is still an underdeveloped research stream in the literature. Much of the research on academic entrepreneurship also is in contrast to industry research, in which early-stage firm interactions have been studied in the context of starting capital, environment, and entrepreneurial orientation (Wiklund and Shepherd 2005). Recent research suggests that the experience of the entrepreneur may be heterogeneous, pointing to the fact there may be successful entrepreneurs who are habitual entrepreneurs (i.e., individuals who are not developing companies for the first time [Westhead, Ucbasaran, and Wright 2005]). Our objective in this study is to focus on characteristics of the academic entrepreneur, to determine whether there are links between industry observations and those within the walls of the university. Additionally, we seek to identify characteristics that correlate to successfully starting a new venture.

The aim of this study is to link levels of commercialization success with characteristics of the academic entrepreneur. In an attempt to better understand how universities might foster new ventures from academic research, we assert that attention should be paid to the mechanisms and attributes that enable the academic entrepreneurial pursuit of successfully created new start-up ventures. As noted by Mosey and Wright (2007), "there is a limited amount of research that has attempted to understand the behavioral and cognitive attributes of entrepreneurial actors and more specifically university scientists who are dedicated to relentless pursuit of commercialization" (Baron 2007; Markman, Siegel, and Wright 2008; Jain, George, and Maltarich 2009). With this in mind, our research initially asks an important question, which is whether "star" academic entrepreneurs have special characteristics (Zucker, Darby, and Brewer 1998). Specifically, our research investigates finer-grained questions regarding academic researchers' experience levels, their motivations, and their ability to foster networks and actively plan to see their technologies commercialized. We explore the entrepreneurial inclination of these academics, both from a quantitative and qualitative perspective, in order to gain insight into their innate characteristics and motivations. In doing so, we undertook a multistep approach beginning with a census of all the invention disclosures over a ten-year period at Northeastern University, a large U.S. research institution located in Boston, Massachusetts. Next, we surveyed the entrepreneurial behaviors of the academic inventors at the same university. Finally, we conducted in-depth interviews of the most successful entrepreneurs to gain firsthand insight into their approach to entrepreneurship within the construct of academic research.

The remainder of the article is organized as follows. The next section outlines the relevant technology transfer and entrepreneurship literature, and positions our study and hypotheses within the space of entrepreneurial characteristics. In section 3, we outline our model and variables. In section 4, we discuss the results. In section 5, we explore the implications of our research and outline policy suggestions. Finally, we conclude with research limitations and directions for further study.

Literature Review and Conceptual Foundations

The study of entrepreneurship in a university context is becoming increasingly important. Academic entrepreneurship, at its most basic level, involves the spillover of research knowledge into commercialized goods and services. Agarwal and colleagues (2007) note that "existing firms and scientists at research institutions facilitate knowledge spillover when they engage in the entrepreneurial act of new venture formation" (267). Over the past thirty years, research universities have functioned as creators and consumers of new knowledge, and their societal role in value creation has become an important policy issue (Markman et al. 2005).

In FY2011, U.S. research universities spent $63.7 billion on research, which resulted in 23,741 invention disclosures; 22,759 patent applications; 40,007 active licenses and options; and a limited number of revenue-generating licensing agreements and university start-ups or spin-outs (AUTM 2013). With respect to the entrepreneurial dimension of technology transfer, the creation of new ventures has been shown to be one of the strongest contributors to innovative activities, competition, economic growth, and job creation (Carree and Thurik 2003). According to a 2004 Venture Impact Study, 10% of the U.S. Gross Domestic Product (GDP) is directly related to new venture creation (Global Insight 2004). A key contribution of this research is that while we examine all of these types of commercialization success, we primarily focus on the latter, the development of a university-based start-up led by an academic entrepreneur.

Even though research suggests that only 12 to 16% of university-assigned inventions are transferred to private new ventures (Di Gregorio and Shane 2003; AUTM 2001 and 2002), university-based start-ups have the greatest potential to affect the greater economy and realize the hopes of government R&D policy makers. New ventures and entrepreneurial activity have been shown to contribute to competition, economic growth, and job creation, and account for a large portion of GDP (Carree and Thurik 2003; Global Insight 2004). Di Gregorio and Shane (2002) found that only a very small percentage of university-assigned inventions are actually transferred or spill over to private new ventures. In this manner, we concur with other scholars that if the purpose of government-funded R&D is to advance the yield from research assets by reducing commercialization cycle time, research on this issue should receive higher levels of consideration (Markman et al. 2005). Shane and Stuart (2002) suggest that start-ups based on university technology such as Genentech, Cirrus Logic, and Lycos tend to survive longer and are more likely to achieve Initial Public Offering (IPO) status. These examples point to the continuing importance of incubation within the university setting (Clarysse et al. 2005).

Di Gregorio and Shane (2002) found that there is substantial evidence that a university's intellectual eminence and licensing policies have a significant impact on start-up activity. They claim that this is because of two reasons: (1) researchers from more prestigious universities are simply better at researching and thus are more likely to create firms to capture the rents to their rare and valuable intellectual property (Zucker, Darby, and Brewer 1998); and (2) inventors from more prestigious universities may be better able to acquire the necessary capital to start their own firms. Considering that these two points are valid, what has yet to be researched are the organizational characteristics of the university (O'Shea et al. 2005), level of funding, age of the TTO, and internal/external resource factors (Powers and McDougall 2005), and more importantly the attributes of the academic inventors themselves. In addition, we argue that it is increasingly important to recognize that firms, universities, and individuals tend to attain only what they are asked to measure as an indicator of their performance. Thus, for true progress to be made there needs to be an accurate assessment of more difficult-to-measure constructs, which involve the entrepreneurial skills of academic entrepreneurs, irrespective of their academic environment, university incentives, or the type of technologies.

There remains a scarcity of "know-why" among university academics and scholars (West 2008) as to how academic research is successfully translated into different types of commercialization success as defined by patents granted, licenses/options received, and formation of university-based start-ups. Accordingly, our goal is to identify some of the characteristics/traits of an academic inventor that affect the commercialization success of their technology or invention. We summarize some of the important papers on university technology transfer, entrepreneurship, and commercialization in table 9.1. We have noted important study variables, research methodology, and key findings.

A majority of the previous work in the area of university technology transfer has focused on the effects of university characteristics (e.g., Agrawal and Henderson 2002; Sherwood and Covin 2008). Along these lines, Baldini, Grimaldi, and Sobrero (2006), highlighted the importance of university size, location, and presence of a medical school on the degree of patenting activity at Italian universities. Powers and McDougall (2005) examined the effect of specific internal and external university resources, such as faculty quality and level of industry R&D funding on the number of start-up companies and new IPOs formed from the university's licensed technology. Louis and colleagues (2001) learned that faculty type, clinical versus nonclinical, had an effect on the likelihood of a patent application being filed or a start-up being formed.

Bozeman (2000) reported that the study of scientific and technical human capital is often underestimated or neglected as criteria for technology transfer effectiveness. The few articles that have looked at characteristics of the academic inventors (individual level) have focused primarily on their effect on patent generation and productivity. Dietz and Bozeman (2005) found that the amount of industry experience of an academic inventor has a positive impact on patent productivity. Renault (2006) found that faculty's own beliefs regarding a university's role in the technology transfer process had a significant impact on their choice to disclose and/or patent their research. Recent research, however, has noted the importance of focusing the lens of empirical research on the academic entrepreneur. An initial attempt at understanding human capital

TABLE 9.1. **Relevant literature for technology transfer, entrepreneurship, and commercialization.**

Indep. Variables	Dep. Variable	References in Literature	Methodology	Results
Academic inventor characteristics	Patents	Allen, Link, Rosenbaum 2007	Survey, 1,335 academic scientists and engineers	Tenured, older, and more experienced males were more likely to patent.
University-level characteristics	Patents	Baldini, Grimaldi, Sobrero 2006	Activities within Italian universities between 1965 and 2002	Positive trend of patents as of 1996. Patent holders were mostly single inventors. Patent activity tripled in universities with IPR regulation.
Resource implications	Spin-outs	Clarysse et. al 2005	In-depth analysis of university spinouts	Examined spinout selection models (low, supportive, and incubator) and resources required for each model (organizational, HR, technological, physical, financial, networking).
Career patterns (diversity, homogeny, education, training)	Patents	Dietz and Bozeman 2005	Secondary data for 1,200 research scientists and engineers in universities	Patent productivity based on total job years in industry and field and cohort differences and center affiliation differences.
Venture capital, intellectual eminence, commercial orientation, licensing policies	Start-ups	Di Gregorio and Shane 2003	Surveyed 101 universities, 457 observations	Intellectual eminence of the university, and the policies of making equity investments in TLO start-ups and maintaining a low inventor share of royalties increased new firm formation activity.
Role identification	Academic entrepreneurship	Jain, George, and Maltarich 2009	Personal and historical semi-structured interviews with 20 faculty, 6 administrators, and 2 technology transfer managers	Due to role identities among scientists, there are multiple understandings of what it means to be "entrepreneurial"(invention disclosure, patents, start-ups).
Collaborative relationships	Research productivity	Landry, Traore, and Godin 1996	Survey of academic researchers from all scientific disciplines	Collaboration, whether within universities, industries, or institutions, increases researchers productivity.

Did you read this Tucker? It is a book.		Libecap 2005		
Secrecy	Marketable knowledge or products	Louis et al. 2001	4,000 clinical and nonclinical life sciences faculty in 49 U.S. research universities	More entrepreneurial faculty (nonclinical) are more likely to be secretive about their research.
UTTO Resources and Competency in identifying licensees, inventor collaboration	Innovation speed (Licensing revenues and new ventures)	Markman et al. 2005	Structured phone interviews with 91 UTTO directors from a sample of 138 U.S. universities	Positive link between commercialization time and licensing revenues and new venture creation. Different resources are needed at various stages of the commercialization process. Early stage, UTTO resources are critical. Late stage, faculty-role is important in innovation speed.
Resources and capabilities, institutional, financial, commercial and human capita	Spin-offs	O'Shea et al. 2005	Panel data from 1980 to 2001	Faculty quality, size, and orientation of science and engineering funding and commercial capability affect university spin-off activity.
University financial, human capital, and organizational resources	Start-up companies and initial public offering which a university had licensed a technology	Powers and McDougall 2005	Data on 120 universities	Industry R&D revenues received by university, quality of their faculty, age of the TTO, and level of venture capital predictor start-ups and IPOs.
University financial, human capital, and organizational resources	Start-up companies and initial public offering which a university had licensed a technology	Powers and McDougall 2005	Data on 120 universities	Industry R&D revenues received by university, quality of their faculty, age of the TTO, and level of venture capital predict start-ups and IPOs.
Institutional policies	Entrepreneurial behavior (collaboration with industry, patenting, and spin-offs)	Renault 2006	Interviews with 98 professors at 12 universities	Most significant influence on entrepreneurial behavior is role of university in the dissemination of knowledge.
External collaborations	Licensing	Thursby and Thursby 2002	Examine three stage licensing factor productivity of 64 universities	Increase of business licensing in from universities is due to increased business reliance on external R&D in faculty research.

and the academic entrepreneur was undertaken by Mosey and Wright (2007). They completed a longitudinal study of human and social capital among academic entrepreneurs and found that business ownership experience and the development of networks and relationships are all important to building successful university spin-outs. Building on this initial stream of work, we seek to fill this gap and identify characteristics of the academic entrepreneur in relation to traditional nonacademic entrepreneurial research. We theorize that there is a direct link between commercialization success and attributes of the academic entrepreneur. We have classified two main areas that build upon existing literature and focus on the academic entrepreneur and commercialization success. These two issues are (1) productivity of the academic entrepreneur and (2) entrepreneurial characteristics of the academic researcher. In the next subsections, we generate our hypotheses.

Academic Entrepreneurs and Productivity

There have been few empirical studies of the translation of research into commercialization success, defined as invention disclosures, patent applications, patents granted, licenses received, and revenue generation via licensing or university-based start-ups. Our agenda is to identify some of the factors that determine an academic inventor's ability to transfer their technology or invention outside the university into a secured license or start-up. Numerous elements could potentially influence the success of an academic inventor's research into industry. O'Shea et al. (2005) found that the amount of government funding had a positive impact on the amount of new spin-off companies generated by a university. Dietz and Bozeman (2005) found that the amount of industry experience of an academic researcher had a positive effect of patent productivity. On the other hand, Powers and McDougall (2005) established that faculty quality (measured by the number of citations) had a positive effect on the number of start-up companies formed and new IPOs. We have identified five different entrepreneurial factors that we believe affect research translation to licensing and spin-outs: (1) academic position (tenure status); (2) participation in an industry sponsored research agreement; (3) research funding source (NSF, etc.); (4) school/college affiliation (engineering versus health sciences, etc.); and (5) research team size. Each one of these factors has an element of productivity that may influence ultimate commercialization success.

Previous researchers have argued that higher-quality faculty members are more interested in the commercialization of their results (Renault 2006). In fact, Renault (2006) found that faculty quality was the most important predictor of technology transfer participation. Academics that actively publish, write research funding proposals, and disclose research can be considered productive. Each one of these areas within research-based science and engineering colleges is a critical criterion for promotion and tenure. A nonproductive faculty member will most likely not be granted tenure. Similarly, O'Shea et al. (2005) found that human capital, specifically faculty quality as measured by the number of post-doctoral staff and faculty, had a positive impact on the number of new spin-off companies generated by a university. This is due to the fact that higher-quality faculty members have greater access to expert knowledge, skills, and talent, and accordingly realize a higher degree of success in their development of cutting-edge technology (O'Shea et al. 2005) and technology transfer efforts (Powers and McDougall 2005). A productive faculty member will seek funding, write and revise proposals, be awarded funds, and coalesce a team around his/her research ideas.

There are a number of factors that can motivate an academic inventor's research including a desire to publish, to achieve tenure, to gain name-recognition (fame), and to gain financial compensation for their work. We assume that academic inventors whose research is driven more by financial motivation will more actively pursue commercialization. Much of the previous work in this area has focused on entrepreneurial motivation at the university level, rather than at the individual academic inventor level. For instance, Baldini, Grimaldi, and Sobrero (2006) found that the adoption of patent regulation at the university level nearly tripled the rate of patenting activities at Italian universities. Thursby and Thursby (2002) found that faculty's degree of entrepreneurship increased both their willingness to disclose and the number of invention disclosures received by the university. Again, a productive academic entrepreneur will seek to disclose and do so repeatedly.

In relating industry entrepreneurial research to the university setting, in an in-depth study of 27 firms over a several-year period, Gartner, Mitchell, and Vesper (1989) found that entrepreneurs who devoted more effort to (1) working with established suppliers or subcontractors,

(2) analyzing potential new entrants, and who (3) were more open to changing the identity of their business, were more likely to start a new venture that survived. The idea that academics who partner with industry gain access to a valuable set of resources and insider knowledge that they can exploit during the technology development and transfer process, we argue, should be very similar to what is seen in entrepreneurial research. The technology transfer literature notes that increased university-industry ties and partnerships result in greater levels of commercialization (O'Shea et al. 2005). Similarly, Blumenthal et al. (1996) found that industry-funded faculty members were more commercially productive than those without industry funding. It has long been argued that financial resources (such as grants from the NSF or NIH) are one of the necessary components to successful technology transfer. Landry, Traore, and Godin (1996) found that collaboration between universities and government increased research productivity.

Greater resources, both financial and human capital, will enable academic inventors to take the necessary steps to seek licenses or form new ventures. In entrepreneurial research, team size has been shown to impact new venture performance. Song, van der Bij, and Halman (2008) found that a larger team may improve new venture performance. In the university setting, this idea has been previously explored by various authors including Markman et al. (2005) who examined research originating from collaborative work of multiple scientists and its association with innovation speed. They hypothesized that collaborative work would enjoy positive spillover effects that would increase the degree of commercialization. Like the work of Song, van der Bij, and Halman (2008) in entrepreneurship, Dietz and Bozeman (2005) found that larger research teams have access to a greater degree of human capital, which in turn impacts research productivity and success.

In sum, an academic entrepreneur who is highly productive in meeting or exceeding the necessary criteria for tenure, developing unique research, obtaining both financial and human capital to leverage those ideas, and disclosing and patenting those ideas should be more successful commercializing university-developed technology. Thus, we hypothesize the following:

H1: An academic entrepreneur's productivity is positively correlated to commercialization success.

Vital Characteristics of Academic Entrepreneurs

Human capital, knowledge, experience, and training in technology transfer is becoming more widely recognized. We assume that human capital is made up of formal education, skills, technical expertise, and experience. Accordingly, several researchers have examined the effect of entrepreneurial or industry experience on commercialization (e.g., Bozeman 2000; Dietz and Bozeman 2005; Louis et al. 2001). Louis and colleagues (2001) examined the effect of entrepreneurial activities by life science faculty on their publishing, teaching, and service productivity. While there is debate on star research and academic entrepreneurs (Van Looy et al. 2004), there is evidence to suggest that habitual academic entrepreneurs can have a substantial impact on new venture creation within the university setting (Mosey and Wright 2007). In this study, we examine four areas that define the academic inventor in terms of their entrepreneurial inclination. The four characteristics we examine are (1) their entrepreneurial experience, (2) their entrepreneurial motivation, (3) their inclination to network, and (4) their focus on commercialization activities and planning.

We conjecture that academic inventors with more entrepreneurial experience have a better understanding of the inner workings of the commercialization process. In essence, through these experiences, academic entrepreneurs accumulate valuable tacit knowledge reservoirs and over time, they are able to reap the benefits associated with their "learning-by-doing" years of expertise and experience (Arrow 1962). Moreover, their experiences in the private sector may have enabled them to develop a unique skill set which will increase the likelihood that their invention or technology will achieve a successful commercial outcome (receive a patent, license, or license option or lead to the formation of a university-based start-up company).

Mosey and Wright (2007) note the importance of developing business networks and relationships with industrial partners. Thus, habitual academic entrepreneurs often seek to develop relationships with individuals outside the academic community for their potential knowledge "spill-in" potential (Mosey and Wright 2007). The idea that academics who partner with external members (industry, government, other institutions) can realize an increase in the in-sourcing or knowledge "spill-in," productivity, and commercial success is not novel and is a well-accepted concept in the knowledge management literature for reducing "core rigidities" (e.g., Leonard-Barton 1992; Markman, Siegel, and Wright 2008). For instance, Landry and colleagues (1996) found that collaborative research increased an academic researcher's productivity. Similarly, Blumenthal and colleagues (1996) found that industry-funded faculty members were more commercially productive than those without industry funding. Finally, Powers and McDougall (2005) found that the level of R&D funding and venture capital investment increased the number of start-ups and new IPOs generated by a university's licensing of technology. Based on this rationale, we argue that academic inventors who engage more actively in sourcing external knowledge "spill-ins" from networking with industry members and building contacts in the private sector will be more likely to have their invention or technology successfully commercialized.

We conjecture that academic inventors who are more commercially motivated or inclined are more likely to have their invention or technology successfully commercialized. These individuals tend to be engaged more in applied research and have a more proactive commercialization plan and commercialization success track record. In support of our argument, Thursby and Thursby (2002) found that changes in faculty research orientation, when faculty research was more oriented toward the needs of business, increased the likelihood of disclosure and subsequent commercialization of academic research. Renault (2006) learned that one of the most important predictors of a faculty member's actual behavior (decision to disclose, patent, and/or form a start-up based on their research) was the individual's beliefs regarding the university's role in commercializing technology.

In sum, we conjecture that an academic's entrepreneurial characteristics play a pivotal role in the commercialization success of university-based research. The keys to success include entrepreneurial experience, motivation, the ability for the academic entrepreneur to actively pursue contacts outside of the university, and the active planning to commercialize the technology. Specifically, we hypothesize the following:

H2: An academic inventor's degree of commercialization success is positively correlated to their entrepreneurial inclination.

In the next section, we review the research method, including the sample, variables, and statistical models.

Methodology

We approached this research as a comprehensive exploratory initiative that triangulates qualitative and quantitative data from databases and controlled surveys to give a concise picture of the academic entrepreneur (Patton 2002). Our technique for data collection and analysis follows the standard hypothetical-deductive method, which includes seven steps. These steps include (1) observation, (2) preliminary information gathering, (3) theory formulation, (4) hypothesizing, (5) further scientific data collection, (6) data analysis, and (7) deduction (Sekaren 1992). Sekaren (1992) noted the importance of this methodology in the study of business and management practices. Siegel, Waldman, and Link (2003) employed these methods in their seminal qualitative analysis on the management practices of university technology transfer offices.

In order to accomplish this task, we analyzed 400 university patent disclosures over a ten-year period at Northeastern University. Next, a survey of these academic entrepreneurs was developed, tested, and implemented. Lastly, we conducted in-depth interviews with eight of the most successful (in terms of gross revenue) academic entrepreneurs that participated in the census and survey. The eight interviews were viewed as distinct cases, following multiple case-study approaches and logic (Yin 1994). Specifics on the sample and data collection are detailed in the next subsection.

Sample/Data

CENSUS OF TTO DISCLOSURES. The data used in our analysis were compiled from a structured analysis of all invention disclosures (IDs) at Northeastern University, a large research (R-1 or R-E) university in Boston, Massachusetts—during a ten-year period (all IDs filed from the inception of the technology transfer office—fiscal year 1999 to fiscal year 2008). Northeastern is home to 37 interdisciplinary research centers and institutes, including an NSF-funded engineering research center and an NSF-funded nanotechnology center. During the study time period, approximately 400 invention disclosures were filed, each corresponding to a unique technology or invention. When filing an invention disclosure form, the discovery or invention is assigned to a primary inventor. These disclosures were filed by 121 different primary academic inventors. Thus, there are individuals in our sample who filed more than one invention disclosure (are identified as the primary inventor on more than one invention disclosure form) during the time period. After eliminating invention disclosure forms with missing or incomplete data, our sample consists of 400 observations; each of the observations correspond to a different technology or invention submitted by an academic entrepreneur on a unique disclosure form. A complete breakdown of the filing dates of the sample is available in tables 9.2a and 9.2b, which contain all of the descriptive statistics of the sample. It is important to note that all revenue generation in the sample is the result of technology-based, new ventures.

The initial stage of the research was conducted in a multistep fashion. The first step was to gather all disclosures at the engaged TTO. Paper files and electronic database records of all invention disclosures were investigated. All pertinent information on disclosure history and current

TABLE 9.2A. **Descriptive statistics of the sample.**

Variable Name	Frequency	Percentage of Sample
Professor	269	67.25%
Assistant/Associate Professor	76	19.00%
Student	15	3.75%
Other	40	10.00%
Sponsored Research Agreement	22	5.50%
Government Funding	103	25.75%
University Funding	21	5.25%
No Funding	114	28.50%
Arts and Science	122	30.50%
Health Sciences	66	16.50%
Computer and Information Science	4	1.00%
Engineering	204	51.00%
Old ID (FY1999–FY2003)	137	34.25%
FY1999	20	5.00%
FY2000	26	6.50%
FY2001	39	9.75%
FY2002	29	7.25%
FY2003	23	5.75%
New ID (FY2004–FY2008)	263	65.75%
FY2004	47	11.75%
FY2005	43	10.75%
FY2006	52	13.00%
FY2007	56	14.00%
FY2008	65	16.25%

TABLE 9.2B. **Descriptive statistics of the sample.**

Variable Name	Mean	Standard Deviation	Median	Minimum	Maximum
Outcome	1.46	0.949	1	1	5
Team Size	2.69	1.496	2	1	9

status were input into a newly created database used in the analysis. Our dependent variable is broken down into five new technological outcomes: (1) the invention or technology is disclosed to the university's TTO (i.e., an invention disclosure form is filed); (2) a non-provisional patent application is filed for the invention or technology;[1] (3) the invention or technology receives a patent; (4) the invention or technology receives a licensing agreement or option; and (5) the invention or technology generates revenue through the licensing option or formation of a university spin-out or new start-up firm. The first category of independent variables applies to the current position or job title of the academic inventor. In this manner, academic inventors/entrepreneurs are separated into four groups depending upon their degree of academic experience: professor, assistant or associate professor, student (this measure encompasses undergraduate and graduate students, lecturers, and postdoctoral researchers), and other (this measure encompasses deans, technical or research specialists, and nonacademic professionals). The position of approximately 67% of the academic inventors included in our sample is listed as professor.

A second category of independent variables deals with whether the academic inventor discovered their invention or technology while participating in an industry-sponsored research agreement (SRA). For approximately 6% of the inventions or technologies included in our sample, the academic inventors discovered their underlying invention or technology as part of a sponsored research agreement. The third category of independent variables pertains to the funding source of the academic inventor's underlying research that was used to discover their invention or technology. In this manner, academic inventors are separated into three groups depending upon the funding source of their research: government funding, university or private funding, and no funding.[2] For approximately 26% of the academic inventors included in our sample, their underlying research or technology was created using some type of government funding. The fourth category of independent variables focuses on the academic inventor's college affiliation within the university. Along these lines, academic inventors are divided according to the colleges within the university in which they are affiliated: the College of Arts and Sciences, the College of Health Sciences, the College of Computer and Information Science, and the College of Engineering. Approximately, 51% of the inventions or technologies included in our sample were filed by academic inventors and were associated with the College of Engineering. The next step of our investigation was analyzing the database via econometrics models.

We employed several types of econometric models owing to their robustness in estimation and their predictive qualities. There are a number of different econometric models that can be used to estimate regressions with ordinal dependent variables. In many cases, researchers use a simple linear regression model. However, unless the dependent variable has a sufficient number of categories, typically assumed to be more than five, this type of regression can produce biased or unreliable results. Some researchers argue that if the dependent variable only has a small number of categories (between three and five), one must estimate either an ordered logit or ordered probit model to ensure the accuracy of the results (Agresti 2002). For our research, we estimated all three models: a simple linear regression, also referred to as an ordinary least squares model (Model 1—OLS Model), an ordered logit model (Model 2), and an ordered probit model (Model 3).[3] The census data was used to evaluate hypothesis 1.

EXPLORATORY EMPIRICAL SURVEY. The survey included questions regarding the academic inventor's position, department, and college affiliation within the university; the type of research they conduct (basic versus applied); their role as an inventor; research team size; funding source; potential commercial applications; research goals and motivation; previous entrepreneurial experience and education; experience with the university TTO; and level of research productivity or success. Thirty-eight academic inventors that were a part of the census data participated in our survey. After eliminating two survey responses with missing or incomplete data, the sample consisted of 36 observations—each corresponding to a unique academic inventor. Post survey, we contacted several of the most successful academic inventors for in-depth interviews. These interviews added significant qualitative insight to our results.

VARIABLES: DEPENDENT AND INDEPENDENT. The dependent variable in our research is a measure of an academic inventor's commercialization success where commercialization success is a measure of how well an academic inventor's technology or invention has been translated into a commercially viable product or knowledge. There are a number of different steps on the path toward commercialization including receiving a patent, obtaining a license or license option, and the forming a start-up. As such, there are four different dependent variables in our analysis—each representing a different type or standard of commercialization outcome. The four different dependent variables used in our analysis include (1) whether a patent was received (Patent Received); (2) whether a license or license option was received (License/Option Received); (3) whether the research/technology led to the formation of a university-based start-up not involving the academic entrepreneur (Start-Up Formed); and (4) whether the academic inventor has started a company related to her/his research (Started Company). All four of the dependent variables are binary variables and only take on the values of zero and one—where a value of zero indicates a failure to achieve this type of commercial success and a value of one indicates a successful attainment of the commercial outcome. Tables 9.3a and 9.3b list the descriptive statistics of the sample.

TABLE 9.3A. **Descriptive statistics of the survey.**

Variable Name	Frequency	Percentage of Sample
Patent Received	22	61.11%
License/Option Received	9	25.00%
Start-Up Formed	10	27.78%
Started Company	11	30.56%
Number of IDs Filed		
1	16	44.44%
2	8	22.22%
3–5	7	19.44%
5–10	2	5.56%
More than 10	3	8.33%
Professor	14	38.89%

TABLE 9.3B. **Descriptive statistics of the survey.**

Variable Name	Mean	Standard Deviation	Median	Minimum	Maximum
Entrepreneurial Experience	1.43	0.677	1.33	0.67	3.00
Entrepreneurial Motivation	2.34	0.703	2.25	1.00	3.81
Networking	2.68	0.719	2.50	1.00	4.00
Commercialization	1.65	0.561	1.67	0.67	2.67

There are four independent variable clusters of interest in our analysis used to evaluate hypothesis 2. The first independent variable cluster is a measure of the academic inventor's entrepreneurial experience. The degree of entrepreneurial experience assigned to the individuals was based upon their responses to three survey questions regarding the extent of their entrepreneurial experience, whether they have ever started a company, and how entrepreneurial they considered themselves. The responses/scores of the three items were then averaged to create a final score. The Cronbach alpha for the scale was 0.72.[4] The scores for entrepreneurial experience in our sample ranged from 0.67 to 3 with an average score of 1.43.

The second independent variable cluster is a measure of the academic inventor's entrepreneurial motivation. The degree of entrepreneurial motivation assigned to an individual was based upon their responses to four survey questions regarding the extent financial gain motivated their research, their desire to obtain licensing revenue, their desire to start a company based upon their research, and their active pursuit of sales and licensing options.[5] The responses/scores of the four items were then averaged to create a final score. The Cronbach alpha for the scale was a 0.58. While the norm for an alpha in a non-exploratory, large sample study is above a 0.70, in many statistical texts an alpha of approximately 0.60 is acceptable for exploratory studies (David 1964; Nunnally 1967). The scores for entrepreneurial motivation in our sample ranged from 1.00 to 3.81 with an average score of 2.34.

The third independent variable cluster is a measure of the academic inventor's networking efforts. The degree of networking assigned to an individual was based upon their response to two survey questions regarding their contribution to networking efforts following the disclosure of their research and the influence of industry members on their decision to patent/patent efforts.[6] The responses/scores to the two items were then averaged to create a final score. The Cronbach alpha for the scale was a 0.70. The scores for networking effort in our sample ranged from 1.00 to 4.00 with an average score of 2.68.

The final independent variable cluster is a measure of the academic inventor's commercialization inclination and efforts. The degree of commercialization assigned to an individual was based upon their responses to four survey questions regarding whether they had a proactive commercial plan for their research, the type of research they perform, their perceived role as a researcher, and the initial intended output of their research.[7] The responses/scores to the four items were then averaged to create a final score. The Cronbach alpha for the scale was 0.57. The scores for commercialization effort in our sample ranged from 0.67 to 2.67 with an average score of 1.65. Scales measures for all four independent variables can be seen in the appendix.

CONTROL VARIABLES. Since some of the academic inventors we surveyed filed more than one invention disclosure with the university's TTO between fiscal years 1999 and 2008, we included a question in our survey pertaining to the number of disclosures each respondent had submitted. Individuals who had submitted more than one disclosure not only had a wider range of research material for which they could realize/obtain commercial success, but may also have more experience with the commercialization process (i.e., a greater familiarity with the patenting, licensing, and start-up formation processes). Therefore, since academic inventors with more inventions or technologies may have a greater/higher probability of having their research obtain a patent, receive a license or option, or lead to the formation of a start-up, we controlled for the number of invention disclosures filed by each respondent. We divided respondents into five categories based upon the number of invention disclosures they had filed.[8] Approximately 56% of the academic inventors included in our sample filed more than one invention disclosure form with the university's TTO between fiscal years 1999 and 2008.

Data Limitations

While we recognize that there are some data limitations in our study owing to the limited size of the sample, we also believe that our qualitative research findings reveal a deeper understanding of the complexity of entrepreneurial behaviors observed in our follow-up analysis. Typically, small samples can also result in inaccurate variances and standard errors, thereby making it difficult to make inferences and reach conclusions regarding the statistical significance of our independent variables of interest. One common technique used to correct for small sample size is the bootstrap resample method, which is used to estimate properties of an estimator (often referred to simply as "bootstrap").[9] Thus, to correct for our small sample size, we use the bootstrap technique along with our model(s) outlined in the next subsection.

Results

The results for the census of the 400 disclosures are shown in table 9.4. As a precaution, robust standard errors were calculated for the linear regression model (OLS Model).[10] Thus, the OLS results listed in tables 9.3a and 9.3b are conservative.

After controlling for disclosure age, we found an academic inventor's research commercialization success is significantly related to a number of different factors. The results of the three models indicate that an academic inventor's position within the university does impact their technological knowledge spillover success. Academic inventors whose job titles were listed as professor were more successful in transferring their technology or knowledge outside the university than academic inventors with lower job titles. This effect is statistically significant at the 1% level in

TABLE 9.4. **OLS, Ordered Logit and Ordered Probit model census results.**

Variables	OLS Model (Model 1) (R2= 0.1105) (N= 400)	Ordered Logit Model (Model 2) (Pseudo R2= 0.0751) (LR Stat = −314.114) (LR Chi-Sq = 50.98) (N= 400)	Ordered Probit Model (Model 3) (Pseudo R2= 0.067) (LR Stat = −316.735) (LR Chi-Sq = 45.74) (N= 400)
Professor	0.286***	0.921***	0.517***
	(0.094)	(0.312)	(0.167)
SRA	0.714***	1.514***	0.804***
	(0.249)	(0.406)	(0.246)
Government Funding	0.400***	0.969***	0.511***
	(0.117)	(0.250)	(0.145)
Engineering	0.045	0.027	0.055
	(0.102)	(0.255)	(0.143)
Team Size	0.023	0.077	0.040
	(0.033)	(0.082)	(0.046)
Old ID	0.170*	0.483*	0.218
	(0.099)	(0.249)	(0.141)
Thresholds			
δ1		2.568	1.460
δ2		3.711	2.087
δ3		4.141	2.305
δ4		5.643	2.978

Standard errors in parentheses
* = 10% significance level ($p < 0.10$), ** = 5% significance level ($p < 0.05$), *** = 1% significance level ($p < 0.01$)

all three models. This result corresponds to the findings of Allen, Link, and Rosenbaum (2007) who found that tenured, older males were more likely to patent. Similarly, the results of the three models also indicate that an academic inventor's participation in an industry-sponsored research agreement affects their degree of technological knowledge spillover success. Academic inventors who had inventions or technologies that were discovered as part of an industry sponsored research agreement were more likely to have their invention or technology be translated into a marketable product, license, or revenue-generating license or spin-out than those who were not part of a sponsored research project. The effect of participation in a sponsored research agreement is statistically significant at the 1% level in all three models. This result corresponds with the findings of Powers and McDougall (2005) that noted the link between industry R&D revenues received and predicting start-ups and IPOs.

Finally, the results of the three models indicate that the funding source of the underlying invention or technology impacts its ability to be successfully transferred outside the university. Academic inventors whose inventions or technologies were discovered using some type of government funding obtained a higher degree of research productivity success than those without government funding. The effect of government funding is statistically significant at the 1% level in all three models.

The results indicate that neither an academic inventor's school/college affiliation within the university nor the size of the research team responsible for discovering the invention or technology appears to affect the degree of research productivity success. The result regarding school/college affiliation may be unique to academic inventors affiliated with the university and therefore may not be true for academic inventors at other universities. Furthermore, we ran an alternate version of the analysis using a subset of our data set only—using just inventions or technologies on invention disclosure forms that were filed more than five years ago (Old IDs Only). We did this in order to address any potential concerns regarding censoring or truncation issues based upon the age of the invention disclosure. As stated previously, inventions or technologies embodied on invention disclosure forms filed more than five years ago would be more likely to be successfully transferred outside the university owing to the lengthy nature of the patenting and licensing processes. The results of these analyses are consistent with the results using the full sample, thus censoring because age of the disclosure does not appear

TABLE 9.5. **Predicted Probabilities from the Ordered Logit model.**

Variable Combination	Invention Disclosure Filed P (Y = 1)	Patent Application Filed P (Y = 2)	Patent Received P (Y = 3)	License or Option Received P (Y = 4)	Revenue Generated P (Y = 5)
Base Observation*	0.914	0.057	0.010	0.015	0.004
Professor	0.809	0.121	0.023	0.036	0.011
Sponsored Research Agreement (SRA)	0.700	0.180	0.039	0.062	0.019
Government Funding	0.801	0.126	0.024	0.038	0.011
College of Engineering	0.912	0.058	0.010	0.015	0.005
All 4 Effects (Professor, SRA, Government Funding, and Engineering)**	0.255	0.263	0.105	0.258	0.119

*Where the base observation is a non-professor, did not participate in a SRA, received no government funding, is not affiliated with the College of Engineering, and has an average research team size.
** Where the four effects observation is a professor, participated in a SRA, received government funding, is affiliated with the College of Engineering, and has an average research team size.

to be an issue. We therefore use our regression results by using the full sample in our discussion below.

Lastly, we looked at predicted probabilities in the Ordered Logit model to investigate the likelihood of obtaining different levels of commercialization success. This is shown in table 9.5.

While the effect of each independent variable alone does not greatly increase the probability that an academic inventor will enjoy a higher degree of commercialization success, when taken in combination they do have a very sizable effect on the ability to successfully transfer technology to the marketplace. According to our results, an academic inventor who is a non-professor with an average-size research team, who disclosed their invention or technology in the past five years, who was unaffiliated with the College of Engineering, with no research funding, and who did not participate in a sponsored research agreement, would have almost no probability of their invention or technology receiving a patent, license, or license option, or having their research generate revenue (0.4% probability). Academic inventors such as these would have a 91% probability that their research would at best be disclosed and only a 6% probability that a patent application would be filed for their invention or technology. Therefore, it would not be very likely that an academic inventor with these characteristics would have their invention or technology be translated into a marketable product via licensing or a university-based spin-off.

On the other hand, an academic inventor who is an experienced professor with an average-size research team, who disclosed their invention or technology less than five years ago, was affiliated with the College of Engineering, had government funding, and who participated in a sponsored research agreement would have a 26% probability of having a patent application filed based upon their research, an 11% probability of having their invention or technology receive a patent, a 26% probability of having their invention or technology receive a license or license option, and a 12% probability that it would generate revenue through a licensing option or the formation of a start-up company. A productive academic entrepreneur who has obtained tenure, actively seeks outside funding, and actively discloses has a substantially higher probability of seeing their idea commercialized. When other factors are included, such as funding type, experience of the faculty member, team size, etc. the probability of commercialization success increases. As such, we find support for hypothesis 1 that there is a correlation between academic entrepreneur productivity and commercialization success. Detailed explanations are discussed in the next section for individual independent variables and combinatorial probabilities. In the next section, the results are discussed in detail along with potential policy implications.

For the exploratory empirical survey of academic inventors, the four outcome models are summarized in table 9.6 (Version A—Single Independent Variable), table 9.7 (Version B—Two Independent Variables), and table 9.8 (Version B—Two Independent Variables). The results presented are for the logit model. The results of the probit model confirm the logit model results. As stated previously, bootstrap standard errors were calculated for all the models. Thus, the results listed in tables 9.6 through 9.8 are conservative. One of the common criteria used to evaluate the effectiveness of discrete choice models is the percent correctly predicted. This "goodness of fit" measure is used to test how well the model fits the data on which it was estimated. The percent correctly predicted is included for each of the models listed in table 9.6, table 9.7, and table 9.8.

In the first step of our analysis, Version A, we included only a single independent variable of interest and our two controls. The results

TABLE 9.6. **Single Independent Variable, Logit Model Results.**

Variables	Model 1 (Patent Received)	Model 2 (License/ Option Received)	Model 3 (Start-Up Formed)	Model 4 (Started a Company)
Entrepreneurial Experience (H1)	−0.187	0.827	1.724**	
	(0.763)	(0.584)	(0.849)	
Number of IDs	1.552**	0.601	0.265	
	(0.667)	(0.373)	(0.415)	
Professor	0.704	1.383	1.450	
	(0.956)	(1.132)	(1.368)	
Percent Correctly Predicted	78.38%	78.38%	75.68%	
Entrepreneurial Motivation (H2)	0.300	0.830	0.316	0.066
	(0.744)	(0.965)	(0.796)	(0.667)
Number of IDs	1.587**	0.639	0.232	−0.062
	(0.758)	(0.495)	(0.487)	(0.472)
Professor	0.719	1.272	0.921	0.778
	(1.162)	(1.073)	(0.955)	(0.828)
Percent Correctly Predicted	81.08%	78.38%	70.27%	67.57%
Networking (H3)	−0.25	0.685	0.081	−0.781
	(0.717)	(0.765)	(0.748)	(0.647)
Number of IDs	1.572**	0.573*	0.231	−0.008
	(0.774)	(0.348)	(0.481)	(0.350)
Professor	0.709	1.250	0.922	0.807
	(0.955)	(1.027)	(1.185)	(0.916)
Percent Correctly Predicted	78.38%	75.68%	70.27%	70.27%
Commercialization (H4)	−0.514	0.300	0.940	1.425*
	(0.664)	(0.763)	(0.945)	(0.730)
Number of IDs	1.606**	0.445	0.076	−0.310
	(0.734)	(0.411)	(0.383)	(0.491)
Professor	0.839	1.377	1.131	1.022
	(1.000)	(0.992)	(1.156)	(0.863)
Percent Correctly Predicted	75.00%	83.33%	72.22%	72.22%

listed in table 9.6 indicate the degree of entrepreneurial experience of the academic inventor has a positive impact on the probability that their research will lead to the formation of a university-based start-up. Academic inventors with a higher degree of entrepreneurial experience were more likely to obtain this type of commercial success than academic inventors with less entrepreneurial experience. This effect is statistically significant at the 5% level. Thus, there is support for hypothesis 2 in the first stage of our analysis (in Model 3, research will lead to the formation of a start-up). We find that academic inventors with a higher degree of entrepreneurial experience are more likely to obtain this type of commercial success than academic inventors with less entrepreneurial experience.

Similarly, the results listed in table 9.6 indicate that the academic inventor's degree of commercialization inclination has a positive impact on the likelihood that they will start a company based on their research. This effect is statistically significant at the 10% level. Thus, there is support for hypothesis 2 in the first stage of our analysis (in Model 4, the academic inventor will start a company). We find that academic inventors with a higher degree of commercialization inclination are more likely to achieve this type of commercial success than academic inventors with less commercialization inclination and activities.

In Model 1, patents awarded is the dependent variable. After controlling for the number of disclosures and faculty position, there is no support for any of our hypotheses and the likelihood that the academic inventor will obtain this type of commercial success. However, the results in Model 1 do confirm our assumption that the number of disclosures positively affects the probability that an academic inventor's invention or technology will receive a patent. On the other hand, in Model 2, where License or Option Received is the dependent variable, while all of the coefficients on the independent variables have the predicted signs, their effects on this type of commercial success are not statistically significant.

Similarly, in Model 3, where the formation of a start-up is the dependent variable, all of the coefficients on the independent variables have the correct/predicted signs. However, other than entrepreneurial experience, none of the other independent variables are statistically significant. After controlling for the number of invention disclosures and faculty positions, there is no support in regard to the likelihood that the academic inventor's invention or technology will lead to the academic inventor to start up a company. Finally, in Model 4, where "the academic entrepreneur started a company" is the dependent variable, after controlling for number of disclosures and faculty quality, there is no support for hypothesis 2. While the coefficient on entrepreneurial motivation has the predicted sign, its effect on this type of commercial success is not statistically significant. On the other hand, the coefficient on networking does not have the correct sign nor is it statistically significant.

In the second step of our analysis, we included a combination of two

TABLE 9.7. **Two Independent Variables, Logit Model Results—Start-Up Formed.**

Variables	Model 3 (Start-Up Formed)
Entrepreneurial Experience	1.613**
	(0.816)
Networking	0.203
	(0.796)
Number of IDs	0.352
	(0.455)
Percent Correctly Predicted	72.97%
Entrepreneurial Experience	1.736*
	(1.030)
Networking	0.286
	(1.194)
Professor	1.598
	(1.661)
Percent Correctly Predicted	75.68%
Entrepreneurial Experience	1.421*
	(0.769)
Commercialization	0.413
	(1.435)
Number of IDs	0.326
	(0.486)
Percent Correctly Predicted	69.44%
Entrepreneurial Experience	1.494*
	(0.903)
Commercialization	0.238
	(1.290)
Professor	1.577
	(1.045)
Percent Correctly Predicted	75.00%

of our independent variables of interest and one of our two controls. This is shown in table 9.7.

After controlling for either the number of disclosures or faculty quality/position in the university, we found limited support for hypothesis 2. The results in table 9.7 indicate after controlling for either the degree of networking or commercialization, the academic inventor's amount of entrepreneurial experience has a positive effect on the probability that their research will lead to the formation of a university based start-up company. Similarly, this effect is statistically significant at the 10% level regardless of whether we control for the number of disclosures or fac-

TABLE 9.8. **Two Independent Variables, Logit Model Results—Started a Company.**

Variables	Model 4 (Started a Company)
Entrepreneurial Motivation	0.566
	(0.689)
Commercialization	1.453*
	(0.778)
Number of IDs	−0.242
	(0.466)
Percent Correctly Predicted	69.44%
Entrepreneurial Motivation	0.031
	(0.800)
Commercialization	1.393*
	(0.831)
Professor	0.873
	(0.953)
Percent Correctly Predicted	72.22%

ulty position. Thus, there is support for hypothesis 2 in the second stage of our analysis (in Model 3, research will lead to the formation of a start-up): academic inventors with a higher degree of entrepreneurial experience are more likely to obtain this type of commercial success than academic inventors with less entrepreneurial experience.

Likewise, the results listed in table 9.8 indicate that even after controlling for the amount of entrepreneurial motivation, the academic inventors' degree of commercialization motivation or inclination has a positive impact on the likelihood that they—the academic entrepreneur—will start a company.

This effect is statistically significant at the 10% level regardless of whether we control for number of disclosures or faculty position. Thus, there is support for hypothesis 2 in the second stage of our analysis (in Model 4, academic inventor will start a company). In the next section, we discuss the results.

Discussion

As a baseline for our study of the antecedents and consequences of academic entrepreneurship, we investigated disclosures and patents, in line with previous studies such as Siegel, Waldman, and Link (2003) and

O'Shea et al. (2005) Our empirical results indicate that an academic inventor who has a higher, tenured position in the university (i.e., whose job title is professor) has a higher probability of having their invention or technology being translated into a marketable product, license, or start-up. We also found that tenured professors are more than twice as likely than non-tenured professors to have patent applications filed for their invention or technology, to have their inventions or technologies receive a patent, have their inventions or technologies receive a license or license option, and to have their work generate revenue through a licensing option or the formation of a university-based start-up company[11] (see table 9.4 for a detailed list of the predicted probabilities from the ordered logit model).[12] Simply stated, to achieve commercialization success the academic entrepreneur needs to be productive as a basis for any ultimate technology transfer success. This result is consistent with others studies such as O'Shea et al. (2005 and 2007). Based on our findings, these academic entrepreneurs need to obtain tenure, have interesting research ideas that can be funded (by industry and government), and need to produce research that is of sufficient interest to be disclosed and patented.

An important result of our study is that academic productivity, as measured by the number of disclosures and resulting patents, is unrelated to the probability that the research will be commercialized. This result indicates that productive academics disclosing research to TTOs and receiving a patent do not have a statistically significant impact on commercialization success. This is important because one of the predominant measures of TTO effectiveness is the generation of disclosures and patents (see Dietz and Bozeman 2005; Allen, Link, and Rosenbaum 2007). In this study, we find that these measures do not reflect whether the research will lead to revenue generation, and points to other factors that may have significance such as experience and networks (Mosey and Wright 2007). Our current research does not assess research quality, but we do find that research quantity in terms of disclosures is no guarantee of commercialization success. In summation, we find that the academic entrepreneur needs to be productive, but that is only a component of the requirements for commercialization success. We next focus on entrepreneurial inclination.

According to Baron (2007), few studies have focused on understanding the behavioral and cognitive attributes of entrepreneurial actors. In a general extension of the initial academic entrepreneur research of Mosey and Wright (2007), the results of our study shed light on some of the important entrepreneurial characteristics of academic inventors that lead to the creation and successful commercialization of new technological innovations. We find that the two most significant characteristics, even after controlling for a number of other characteristics and environmental conditions, are the academic inventor's entrepreneurial experience and his/her commercialization activities or inclination toward commercializing his/her research. These two characteristics alone significantly account for why some academic inventors are more successful than others in transferring their research outside the university into the private sector through a new, start-up venture.

A main contribution of this research is linking entrepreneurial qualities and the formation of new ventures. In our sample, the percentage of academic entrepreneurs who have started a company is high. Some 30% of the Northeastern professors studied have started a company based on their research, which is more than twice the rate found in other studies (Di Gregorio and Shane 2003). As such, this represents an interesting population to explore our hypotheses on specially, entrepreneurial inclination. What we found is telling for universities that are interested in converting research, both basic and applied, into revenue-producing spin-outs. We found support for hypothesis 2, which was that academic researchers with entrepreneurial experience are more likely to see their research commercialized via a start-up. Characteristics of these successful academic entrepreneurs included previous experience in starting a company, prior education in entrepreneurship, and a belief that they were entrepreneurs—not solely academics. These academic entrepreneurs saw commercialization as an integral stage in the maturation of their research. In one of our follow-on, in-depth interviews, one senior academic entrepreneur stated that "I've started several companies before—I view beginning a new venture as a logical next step in the maturation of my research."

Our results show that academics at Northeastern with a stronger inclination to commercialize their research were more likely to start their own companies. Our research results also indicate that an academic inventor's participation in an industry-sponsored research agreement has a positive effect on his/her commercialization success, relating to their stronger inclination to commercialize. Specifically, we found that an academic inventor who participated in a sponsored research agreement was more than three times as likely as a nonparticipant to have a patent application filed for their invention or technology and to receive a patent. Similarly, an academic inventor who participated in a sponsored research agreement was more than four times as likely than a nonparticipant to have their research receive a license and to have their work generate revenue through a licensing option or the formation of a new venture.[13] This finding is consistent with the findings of previous research regarding the positive effect of harnessing *knowledge spill-ins* from external industry leaders (Agarwal, Audretsch, and Sarkar 2007; Dietz and Bozeman 2005; O'Shea et al. 2005; and Powers and McDougall 2005).

These entrepreneurs strategically planned for commercial outlets from the onset of research and tended to seek targeted (applied) projects that have a clear horizon for application. They also viewed commercialization as a critical phase in the research process. From one of our qualitative interviews, we were better able to understand these phenomena. One of our respondents was a highly focused, financially driven biotechnology professor. He explained that his highly successful spin-out is a result of the fact that he was "in total control of the project to ensure its commercialization and was very financially motivated and wanted his research to be a success."

Such findings suggest a number of interesting practical implications. First, our results have shown that academic entrepreneurs need to be productive within their environment to have commercialization success. Secondly, we show that being an entrepreneur is a key attribute of commercialization success. We found that faculty with more entrepreneurial experiences were more likely to have their invention or technology translated into a university based start-up. Interestingly, these empirical and qualitative follow-up findings imply if universities are interested in bolstering research-based new ventures, universities should try to actively recruit faculty with more entrepreneurial experience. We define entrepreneurial experience as individuals that have started their own company or worked at a start-up, and view themselves to be entrepreneurs in the traditional sense, not just *entrepreneurial*. Jain and colleagues (2009) found academic scientists tend to have different definitions of what it means to be entrepreneurial (disclosing inventions, filing patents, etc.). In seeking academic entrepreneurs—those experienced in the start-up world and inclined to see new ventures as the culmination of their research—universities may need to broaden the scope of activities from recruiting requirements to rewards and promotion.

In an effort to increase positive reinforcement, responsibility, and rewards for academic entrepreneurship, universities should also look at incorporating entrepreneurial achievements, such as inventing, patenting, licensing, and creating spin-offs, into their promotion and tenure policies. In essence, universities, like all businesses, need to recognize that all too often "you get what you reinforce" (Luthans and Stajkovic 1999). Thus, it may be advisable for universities to create and promote entrepreneurial cultures on their campuses that help to prevent the siphoning of faculty members with entrepreneurial experience to the private sector. Along these same lines, universities may need to adopt policies that promote entrepreneurial thinking and learning. This could include such activities as lecture series and seminars on recent entrepreneurial issues and concerns, hands-on training and education in the development of entrepreneurial thinking. Bercovitz and Feldman (2008) note that a bottom-up approach to changing the university organization to foster a climate more conducive to the academic entrepreneur and associated technology transfer initiatives may be preferred.

A successful example that contributes to these kinds of entrepreneurial faculty-friendly policies is the Massachusetts Institute of Technology (MIT). The cofounder of A123Systems, a successful university spin-out, was encouraged to take a leave of absence at MIT to lead an outside research team that contributed to the development of new battery technology (Bowen, Morse, and Cannon 2006). In this case, not only did the lead researcher at MIT have the entrepreneurial experience necessary to create such a success story, but he also worked in a university that fostered entrepreneurship and industry collaboration. Overall, our findings suggest that academic inventors that apply the lessons learned from these types of successful experiences are able to increase their speed and innovative abilities when faced with similar opportunities. In essence, innovation builds upon existing knowledge domains (e.g., Nerkar 2003).

Finally, while it is important to note that universities can and should engage in these and a number of other strategies to promote the growth of entrepreneurial experience and commercialization among their faculty, they are not alone in the struggle to create greater academic entrepreneurship. The federal government and other private funding agencies need to also seek out faculty with strong academic entrepreneurial characteristics that have a successful record of accomplishment of transferring knowledge outside of the university in the form of successful new, start-up companies. Sponsoring more applied research can be a strategy to accomplish this task (e.g., funding from the NIH, DARPA, NSF, etc.). As one successful academic entrepreneur succinctly stated in our in-depth interviews, "Our research was funded by DARPA. We had a clear path towards commercialization." In sum, university scientists have to have specific "know why" characteristics (West 2008), such as productivity, entrepreneurial experience, and inclination, as well as internal and external support and funding. This combination will help enhance the probability of success of their commercialization activities. The importance of these leading innovation activities is profound, as they have a cascading effect and can further act as pistons in the engine of economic growth and job creation in the U.S. (Agarwal, Audretsch, and Sarkar 2007).

Limitations, Future Research, and Conclusion

A university's academic eminence, equity involvement in their start-ups, and royalty rewards to faculty may explain some of the variation in technology transfer outcomes (Di Gregorio and Shane 2003). A limitation of our study is that we were unable to investigate academic entrepreneurial characteristics across multiple universities (Markman et al. 2005). Given the limited number of research studies that have examined the behavioral attributes of entrepreneurial actors (Baron 2007), particularly among university scientists (Jain, George, and Maltarich 2009), we suggest that future researchers may want to examine our exploratory model with a larger sample size. In spite of these limitations, our findings are novel and suggest that entrepreneurial productivity, experience, and commercialization inclination play a critical role in whether or not a research initiative is commercially successful outside of the academic institutional setting.

Epilogue

After completing our research, Northeastern replaced its traditional technology transfer office with the Center for Research Innovation (CRI) (http://www.northeastern.edu/research/cri/). In an effort to bridge the gap between laboratory research and need-based solutions, CRI was created. The CRI serves as the university-wide portal between industry and leading-edge innovations from Northeastern's use-inspired research portfolio. A major focus of the new CRI initiative is to foster campus entrepreneurship between research faculty, students, and industry. A majority of the CRI staff has prior entrepreneurial experience, creating an agile and responsive team focused on the translation of university innovations into tangible solutions through licenses, spin-outs and collaboration. Operationally, CRI sponsors Global Entrepreneurship Week and yearly research expositions, which highlight both student and faculty research. This is designed as a more proactive and entrepreneurial approach to technology transfer, licensing, and spin-outs.

Appendix

TABLE 9A.1. **Scale Measures.**

Entrepreneurial Experience
Q1: 5-point Likert
Please rate your previous entrepreneurial experience:
1. No prior entrepreneurial experience
5. A great deal of prior entrepreneurial experience
Q2: Binary
Have you ever started a company?
0. No
1. Yes
Q3: 5-point Likert
How entrepreneurial do you consider yourself?
1. I'm an academic researcher, not an entrepreneur
2. Mostly research, little entrepreneurial skill
3. Good balance of academic and entrepreneurial skills
4. Very entrepreneurial
5. Extremely entrepreneurial, a primary goal of mine
Cronbach Alpha = 0.72

Entrepreneurial Motivation
Q1: 5-point Likert
Please rate the level of financial motivation pertaining to your research:
1. No motivation
5. Extremely motivated to see financial rewards as a result of my research
Q2: 5-point Likert
Please rate your motivation to obtain licensing revenue from your research:
1. No motivation to see my research licensed
5. Extremely motivated to see my research licensed
Q3: 5-point Likert
Please rate your motivation to start a company based on your research:
1. No motivation to start a company based on my research
5. Extremely motivated to start a company based on my research
Q4: 5-point Likert
Please rate how actively you pursued sales and licensing opportunities:
1. No effort in pursuing sales and licensing opportunities
5. Extremely active in pursuing sales and licensing opportunities
Cronbach Alpha = 0.58

(continued)

TABLE 9A.1. *(continued)*

Networking Characteristics
Q1: 5-point Likert
Please rate your contribution to networking your research post disclosure:
1. No networking was performed post disclosure
5. Networking was pervasive post disclosure
Q2: 5-point Likert
Please rate the influence outside industry contacts has on you pursing commercialization:
1. No influence in pursuing commercialization of my research
5. Extremely influential in pursuing commercialization of my research
Cronbach Alpha = 0.70

Commercialization Inclination and Activities
Q1: Binary scale
Please rate the extent to which you actively developed a commercialization plan:
0. No commercialization plan actively developed
1. Actively developed a commercialization plan
Q2: 5-point continuum
Please select the type of research performed:
1. Basic
5. Applied
Q3: 5-point Likert
Please evaluate your role as a researcher:
1. I perform research, commercialization is another department's job
2. I'll advise on the project, but it's not my main responsibility
3. I'll be a part of a team, but not the lead
4. I'll be a project leader, but will need support from other departments
5. I want to be in total control to ensure commercialization
Q4: 3-point scale
What was the primary intended output of the research?
I want to be in total control to ensure commercialization
1. Publications only
2. Develop a patentable discovery
3. License the technology
Cronbach Alpha = 0.57

Notes

1. There are a wide variety of patents, which can be applied for including provisional and non-provisional patents. We take a more restrictive approach toward what qualifies as a patent application, and thus we only include patent applications for non-provisional patents.

2. For a limited number of invention disclosures the funding source was unavailable. In order to preserve the sample size, the funding source for these invention disclosures is identified simply as unknown.

3. We estimated a simple linear model merely as an exploratory analysis and as an additional robustness check for our results.

4. A Cronbach alpha is an unbiased estimator of reliability of the scale created from all the independent variables included in a cluster. The alpha computes the correlations and covariances for all pairs of variables included in the

cluster. An alpha of above 0.70 is the norm for non-exploratory, large sample studies.

5. In order to preserve the sample size, respondents who left any of these questions blank used to calculate the degree of entrepreneurial motivation were replaced with the sample average for the question rather than assigning a missing value.

6. In order to preserve the sample size, respondents who left any of these questions blank used to calculate the degree of networking were replaced with the sample average for the question rather than assigning a missing value.

7. In order to preserve the sample size, respondents who left any of these questions blank used to calculate the degree of commercialization were replaced with the sample average for the question rather than assigning a missing value.

8. Our sample can be broken down as follows: 16 of our respondents filed one invention disclosure; 8 respondents filed two invention disclosures; 7 respondents filed between three and five invention disclosures; 2 respondents filed between five and ten invention disclosures; and 3 respondents filed greater than ten invention disclosures.

9. The bootstrap method was invented by Bradley Efron and further developed by Efron and Tibshirani (Efron 2000).

10. There is not much difference between the regular and robust standard errors in our OLS model. Thus our model does not appear to suffer from heteroskedasticity or misspecification of the error structure.

11. Non-professors have a 5.7% probability of having a patent application filed for their invention or technology, a 1% probability that it will receive a patent, a 1.5% probability that it will receive a license or license option, and less than a 1% probability that it will generate revenue. On the other hand, professors have a 12.1% probability of having a patent application filed for their invention or technology, a 2.3% probability that it will receive a patent, a 3.6% probability that it will receive a license or license option, and greater than a 1% probability that it will generate revenue.

12. See table 6 for the predicted probabilities from the ordered logit model using a subsample of the data just invention disclosures that were filed more than five years ago (Old IDs Only).

13. Non-participants have a 5.7% probability of having a patent application filed for their invention or technology, a 1% probability that it will receive a patent, a 1.5% probability that it will receive a license or license option, and less than a 1% probability that it will generate revenue. On the other hand, participants have an 18% probability of having a patent application filed for their invention or technology, a 3.9% probability that it will receive a patent, a 6.2% probability that it will receive a license or license option, and slightly less than a 2% probability that it will generate revenue.

References

Adams, J. 1990. "Fundamental stocks of knowledge and productivity growth." *Journal of Political Economy* 98: 673–702.

Agrawal, A. and R. Henderson. 2002. "Putting patents in context: Exploring knowledge transfer from MIT." *Management Science* 48 (1): 44–60.

Agarwal, R., D. Audretsch, and M. B. Sarkar. 2007. "The process of creative construction: Knowledge spillovers, entrepreneurship and economic growth." *Strategic Entrepreneurship Journal* 1: 263–86.

Agresti, A. 2002. *Categorical Data Analysis, in Probability and Statistics*. New York: Wiley and Sons.

Allen, S. D., A. N. Link, and D. T. Rosenbaum. 2007. "Entrepreneurship and human capital: Evidence of patenting activity from the academic sector." *Entrepreneurship Theory and Practice* 31 (6): 937–51.

Arrow, K. 1962. "The economic implications of learning by doing." *Review of Economic Studies* 29: 155–73.

AUTM (Association of University Technology Managers). *Surveys FY1991 to FY2009*. http://www.autm.net/Surveys.htm. Accessed May 10, 2011.

AUTM, AUTM US FY12. "Highlights: Association of University Technology Managers." http://www.autm.net/. Accessed October 28, 2013.

Baldini, N., R. Grimaldi, and M. Sobrero. 2006. "Institutional changes and the commercialization of academic knowledge: A study of Italian universities' patenting activities between 1965 and 2002." *Research Policy* 35: 518–32.

Baron, R. A. 2007. "Behavioral and cognitive factors in entrepreneurship: Entrepreneurs as the active element in new venture creation." *Strategic Entrepreneurship Journal* 1: 167–82.

Bercovitz, J., and M. Feldman. 2008. "Academic Entrepreneurs: Organizational Change at the Individual Level." *Organization Science* 19 (1): 69–89.

Blumenthal, D., et al. 1996. "Participation of life-science faculty in research relationships with industry." *New England Journal of Medicine* 335 (23): 1734–39.

Bowen, H. K., K. P. Morse, and D. Cannon. 2006. A123Systems: *Harvard Business School Case* (May 8): 9-606-114.

Bozeman, B. 2000. "Technology transfer and public policy: A review of research and theory." *Research Policy* 29: 627–55.

Carree, M. A., and A. R. Thurik. 2003. "The Impact of Entrepreneurship on Economic Growth." In *Handbook of Entrepreneurship Research*, edited by D. B. Audretsch and Z. J. Acs, 437–71. Boston: Kluwer.

Clarysse, B., et al. 2005. "Spinning out new ventures: A typology of incubation strategies from European research institutions." *Journal of Business Venturing* 20: 183–216.

David, F. B., 1964. *Educational Measurements and Their Interpretation*. Belmont, CA: Wadsworth.

Dietz, J. S., and B. Bozeman. 2005. "Academic careers, patents, and productivity: Industry experience as scientific and technical human capital." *Research Policy* 34: 349–67.

Di Gregorio, D., and S. Shane. 2003. "Why do some universities generate more start-ups than others?" *Research Policy* 32: 209–27.

Efron, B. 2000. "The bootstrap and modern statistics." *Journal of the American Statistical Association* 95: 1293–96.

Efron, B., and R. Tibshirani. 2000. *An Introduction to the Bootstrap (Monographs on Statistics and Applied Probability).* New York: Chapman and Hall.

Gartner, W. B., T. R. Mitchell, and K. H. Vesper. 1989. "A taxonomy of new business ventures." *Journal of Business Venturing* 4 (3): 169–86.

Global Insight. 2004. "Venture Impact 2004: Venture Captial Benefits to the U.S. Economy." White paper prepared for the National Venture Capital Association. Arlington, VA.

Jain, S., G. George, and M. Maltarich. 2009. "Academics or entrepreneurs? Investigating role identify modification of university scientists involved in commercialization activity." *Research Policy* 38: 922–35.

Kanuk, L. and C. Berenson. 1975. "Mail surveys and response rate: A literature review." *Journal of Marketing Research* 22: 440–53.

Landry, R., N. Traore, and B. Godin. 1996. "An econometric analysis of the effect of collaboration on academic research productivity." *Higher Education* 32: 283–301.

Libecap, G.D. ed. 2005. *University Entrepreneurship and Technology Transfer: Process, Design, and Intellectual Property.* San Diego, CA: Elsevier, Ltd.

Leonard-Barton, D. 1992. "Core capabilities and core rigidities: A paradox in managing new product development." *Strategic Management Journal* 13: 11–125.

Lockett, A., et al. 2005. "The creation of spin-off firms at public research institutions: Managerial and policy implications." *Research Policy* 34 (7): 981–93.

Louis, K. S., et al. 2001. "Entrepreneurship, secrecy, and productivity: A comparison of clinical and non-clinical life science faculty." *Journal of Technology Transfer* 26: 233–45.

Luthans, F., and A. D. Stajkovic. 1999. "Reinforce for performance: The need to go beyond pay and even rewards." *Academy of Management Executive* 13 (2): 49–57.

Markman, G. D., et al. 2005. "Innovation speed: Transferring university technology to market." *Research Policy* 34: 1058–75.

Markman, G. D., D. S. Siegel, and M. Wright. 2008. "Research and technology commercialization." *Journal of Management Studies* 45(8): 1410–23.

Mosey, S., and M. Wright. 2007. "From Human Capital to Social Capital: A Longitudinal Study of Technology-Based Academic Entrepreneurs." *Entrepreneurship Theory and Practice* 31 (6): 909–35.

Mowery, D. C. 2004. "The Bayh-Dole Act and High Technology Entrepreneurship in U.S. Universities: Chicken, Egg, or Something Else?" *University Entrepreneurship Technology Transfer: Process, Design, and Intellectual Property Advances in the Study of Entrepreneurship, Innovation and Economic Growth* 16: 39–38.

Nerkar, A. 2003. "Old is gold? The value of temporal exploration in the creation of new knowledge." *Management Science* 49 (2): 211–29.

Nunnally, J. C. 1967. *Psychometric Theory*. New York: McGraw-Hill.

O'Shea, R. P., et al. 2005. "Entrepreneurial Orientation, Technology Transfer and Spinoff Performance of U.S. Universities." *Research Policy* 34: 994–1009.

O'Shea, R. P., et al. 2007. "Delineating the anatomy of an entrepreneurial university: The Massachusetts Institute of Technology experience." *R&D Management* 37: 1–16.

Oppenheim, A. N. 1966. *Questionnaire Design and Attitude Measurement*. New York: Free Press.

Patton, M. Q. 2002. *Qualitative Research and Evaluation Methods*. 3rd ed. Thousand Oaks, CA: Sage.

Powers, J. B., and P. P. McDougall. 2005. "University start-up formation and technology licensing with firms that go public: A resource-based view of academic entrepreneurship." *Journal of Business Venturing* 20: 291–311.

Renault, C. S. 2006. "Academic Capitalism and University Incentives for Faculty Entrepreneurship." *Journal of Technology Transfer* 31: 227–39.

Sekaren, U. 1992. *Research Methods for Business: A Skill Building Approach*. 2nd ed. New York: John Wiley and Sons.

Shane S., and T. Stuart. 2002. "Organizational Endowments and the Performance of University Start-Ups." *Management Science* 48 (1): 154–70.

Sherwood, A. L., and J. G. Covin. 2008. "Knowledge acquisition in university-industry alliances: An empirical investigation from a learning theory perspective." *Journal of Product Innovation Management* 25 (2): 162–79.

Siegel, D. S., and P. H. Phan. 2004. "Analyzing the Effectiveness of University Technology Transfer: Implications for Entrepreneurship Education." *University Entrepreneurship Technology Transfer: Process, Design, and Intellectual Property Advances in the Study of Entrepreneurship, Innovation and Economic Growth* 16: 1–38.

Siegel, D. S., D. Waldman, and A. N. Link. 2003. "Assessing the Impact of Organizational Practices on the Relative Productivity of University Technology Transfer Offices: An Exploratory Study." *Research Policy* 32 (1): 27–48.

Song, M. K. Podoynitsyna, H. van der Bij, and J. Halman. 2008. "Success factors in new ventures: A meta-analysis." *Journal of Product Innovation Management* 25 (1): 7–27.

Thursby, J. G., and M. C. Thursby. 2002. "Who is selling the ivory tower? Sources of growth in university licensing." *Management Science* 48 (1): 90–104.

Van Looy, B., et al. 2004. "Combining entrepreneurial and scientific performance in academia: Towards a compounded and reciprocal Matthew-effect?" *Research Policy* 33 (3): 425–41.

West, J. 2008. "Commercializing open science: Deep space communications as the lead market for Shannon Theory, 1960–73." *Journal of Management Studies* 45: 1506–32.

Westhead, P., D. Ucbasaran, and M. Wright. 2005. "Decisions, Actions, and Performance: Do Novice, Serial, and Portfolio Entrepreneurs Differ?" *Journal of Small Business Management* 43 (3): 393–417.

Wiklund, J., and D. Shepherd. 2005. "Entrepreneurial orientation and small business performance: A configurational approach." *Journal of Business Venturing* 20 (1): 71–91.

Yin, R. K. 1994. *Case Study Research: Design and Methods*, Thousand Oaks: Sage Publications.

Zhao, L. M., and A. Reisman. 1992. "Toward meta research on technology-transfer." *IEEE Transactions on Engineering Management* 39 (1): 13–21.

Zucker, L.G., M. R. Darby, and M. B. Brewer. 1998. "Intellectual Human Capital and the Birth of U.S. Biotechnology Enterprises." *American Economic Review* 88 (1): 290–306.

Contributors

David Audretsch
Institute of Development Strategies
Indiana University
Bloomington, IN 47405
U.S.

Nicola Baldini
Department of Management
University of Bologna
Bologna, Italy 40126

Thomas Byers
Department of Management Science and Engineering
Stanford University
Stanford, CA 94305
U.S.

Denise Dunlap
College of Business Administration
Northeastern University
Boston, MA 02115
U.S.

Riccardo Fini
Department of Management
University of Bologna
Bologna, Italy 40131
and
Imperial College London
Business School
United Kingdom

John H. Friar
College of Business Administration
Northeastern University
Boston, MA 02115
U.S.

Devrim Göktepe-Hultén
Research Policy Institute Lund University School of Economics and Management
Lund, Sweden

Rosa Grimaldi
Department of Management
University of Bologna
Bologna, Italy 40126

Julian Kolev
SMU Cox School of Business
Dallas, TX 75275
U.S.

Albert N. Link
Department of Economics
University of North Carolina at Greensboro
Greensboro, NC 27402
U.S.

Tucker J. Marion
College of Business Administration
Northeastern University
Boston, MA 02115
U.S.

Fiona Murray
MIT Sloan School of Management
Cambridge, MA 02142
U.S.

Andrew Nelson
Lundquist College of Business
University of Oregon
Eugene, OR 97405
U.S.

Markus Perkmann
Business School
Imperial College London
London SW7 2AZ
U.K.

Arti Rai
Duke Law School and Duke Institute for Genome Sciences and Policy
Durham, NC 27708
U.S.

Bhaven Sampat
Department of Health Policy and Management Mailman School of Public Health
Columbia University
New York, NY 10032
U.S.

John T. Scott
Department of Economics
Dartmouth College
Hanover, NH 03755
U.S.

Donald S. Siegel
School of Business
University at Albany, SUNY
Albany, NY 12222
U.S.

Joel West
Keck Graduate Institute
The Claremont Colleges
Claremont, CA 91711
U.S.

Mike Wright
Imperial College Business School
London SW72AZ
U.K.

Index